The elephant and the dragon in contemporary life sciences

Manchester University Press

INSCRIPTIONS

Series editors
Des Fitzgerald and Amy Hinterberger

Editorial advisory board
Vivette García Deister, National Autonomous University of Mexico
John Gardner, Monash University, Australia
Maja Horst, Technical University of Denmark
Robert Kirk, Manchester, UK
Stéphanie Loyd, Laval University, Canada
Alice Mah, Warwick University, UK
Deboleena Roy, Emory University, USA
Hallam Stevens, Nanyang Technological University, Singapore
Niki Vermeulen, Edinburgh, UK
Megan Warin, Adelaide University, Australia
Malte Ziewitz, Cornell University, USA

Since the very earliest studies of scientific communities, we have known that texts and worlds are bound together. One of the most important ways to stabilise, organise and grow a laboratory, a group of scholars, even an entire intellectual community, is to write things down. As for science, so for the social studies of science: Inscriptions is a space for writing, recording and inscribing the most exciting current work in sociological and anthropological – and any related – studies of science.

The series foregrounds theoretically innovative and empirically rich interdisciplinary work that is emerging in the UK and internationally. It is self-consciously hospitable in terms of its approach to discipline (all areas of social sciences are considered), topic (we are interested in all scientific objects, including biomedical objects) and scale (books will include both fine-grained case studies and broad accounts of scientific cultures).

For readers, the series signals a new generation of scholarship captured in monograph form – tracking and analysing how science moves through our societies, cultures and lives. Employing innovative methodologies for investigating changing worlds, it is home to compelling new accounts of how science, technology, biomedicine and the environment translate and transform our social lives.

Previously published titles

Trust in the system: Research Ethics Committees and the regulation of biomedical research   Adam Hedgecoe

Embodiment and everyday cyborgs: Technologies that alter subjectivity
Gill Haddow

Personalised cancer medicine: Future crafting in the genomic era
Anne Kerr *et al.*

The elephant and the dragon in contemporary life sciences

A call for decolonising global governance

Joy Y. Zhang and Saheli Datta Burton

MANCHESTER UNIVERSITY PRESS

Published by Manchester University Press
Oxford Road, Manchester M13 9PL
www.manchesteruniversitypress.co.uk

British Library Cataloguing-in-Publication Data
A catalogue record for this book is available from the British Library

ISBN 978 1 5261 5952 6 hardback
ISBN 978 1 5261 8228 9 paperback

First published 2022

The publisher has no responsibility for the persistence or accuracy of URLs for any external or third-party internet websites referred to in this book, and does not guarantee that any content on such websites is, or will remain, accurate or appropriate.

Typeset
by New Best-set Typesetters Ltd

Contents

Preface

This book discusses how the global governance of the life sciences can be improved in the face of emerging powers. The rise of India's and China's life sciences are used as examples. We use the terms 'elephant' and 'dragon' metaphorically. That is, China and India are *both* seen as emerging 'dragons' *and* as 'elephants'. Both countries have formidable resources and are boldly determined to have their presence felt. Both have been increasingly recognised as significant partners at the international science forums. Yet, even when regulatory pledges are made between Western institutions and China or India during international meetings, there often remains an 'elephant in the room'. Would these two Global South scientific 'dragons' really abide by the agreed rules? Regulatory enforcement has been a chronic issue for both countries. This is not simply an issue of underdeveloped institutional structures or a lagging behind in managerial techniques. To borrow a concept from theorist Walter D. Mignolo (2009), the bigger 'elephant' – the apparent yet unsettling issue – is that the two emerging powers are 'epistemically disobedient'.

Such 'disobedience' is rooted in both passive and active considerations. At one level, cultural and developmental differences have pressured the two countries to have different scientific priorities, and forge different strategies in the funding, organisation and delivery of research. These context dependent deviations from mainstream (Western) science aside, at another level, there is an active effort in challenging existing global epistemic hierarchies in defining what 'good science' is, how it is to be validated and who constitutes credible knowers (Zhang, 2012a). The socio-economic and political impact of science are purposefully used by both states and by socio-scientific communities within the two emerging countries to restructure the pecking order of global knowledge systems and by extension, to reshape the political landscape both within and beyond their national boundaries. It is also important to be reminded that, at least at the state level, neither China nor India are in complete control of how science is developed. With the proliferation of civic funding, the popularisation (and to some extent, deskilling) of technical know-how and the increasingly transnational and trans-disciplinary geography of research projects, even authoritarian regimes such as China find it difficult to track every new scientific initiative, let alone to identify and effectively respond to their socio-ethical consequences.

In this sense, the rise of the life sciences in China and India has a *magnifying glass effect* on some of the fundamental challenges that are shared by scientific communities worldwide. As demonstrated in this book, while some of the governing challenges faced by the two countries are local, it would be a mistake to simply dismiss their experiences as the 'Other', or worse, to assume the 'elephant' and the 'dragon' would be better disciplined if only they could act more like the West. Arguably what makes India and China a worry for bioethicists and regulators is the fact that both countries are presenting scenarios that neither the West nor anyone else has answers to. The 'subversiveness' they present and represent mirror a wider phenomenon in which previously marginalised (networks of) actors, both from the Global South and Global North, contest the conventional thinking of how science and scientists should be governed (e.g. Ong and Chen, 2010; Petryna, 2009; Rosemann and Chaisinthop, 2016;). As the trajectory of contemporary science has eroded the boundaries of traditional colonies of expertise and authority, few would disagree that good governance necessarily needs to think 'outside of the box', and needs to be 'de-colonised' as well.

To take governing challenges seriously, we need to bring new wisdom into the old saying that 'science is universal'. Is science 'universal'? Conventionally, from the perspective of the Global South, such a claim itself is an assertion of Eurocentric epistemologies. For science has never been an exercise of disinterested fact gathering, but has always been 'an uneven distribution of epistemic authority', which imposes order over knowledge and those who know (Hamilton, Subramaniam and Willey, 2017: 615). But in terms of its *impact*, we argue that the global outreach of science cannot be emphasised enough. As political scientist Brian Salter (2008: 156) rightly argued after examining stem cell governance in both China and India, in contemporary global innovation, 'no state can afford to be a political island'. Science governance would be much easier if only the social-ethical consequences of controversial practices or disputable risk-benefit judgements could be confined to particular geographical boundaries, or to a neat West-East divide (Bharadwaj, 2012). But as we have seen in the life sciences, from experimental trials of regenerative medicine to reproductive human genome editing, the exploitation of scientific viability in one corner of the world often profoundly re-orients the perception of technological imperatives and associated risk calculations around the world. Thus, to maximise the benefits and to minimise the risk of emerging science in aiding our collective pursuit of a good life, we need to acknowledge and effectively respond to the universal potential of science's impact. For both contemporary science and its governance, we need what Sandra Mitchell (2009: 16) called 'integrative pluralism', in which a plurality of perspectives can be integrated to reflect the dynamic and evolving character of knowledge.

This book does not argue for a simplistic expansion of scope in Western discussions to include China and India as 'alternatives'. Rather, it argues that a more radical shift in our global vision of science governance is needed and that such a shift requires mutual understanding and coordination from both the developed and developing worlds.

Our discussion builds on and extends an increasing number of decolonial examinations of science (e.g. Chakrabarty, 2000; Harding, 1998, 2008; Santos, 2014; Santos and Meneses, 2020). We demonstrate that the epistemic disobediences exhibited in India and China are *constitutive* of contemporary science. Until recently, the critical gaze on emerging scientific capacity in the Global South was cast on the 'diffusion' of norms from the centre to the periphery (i.e. the West to the Rest). There are good reasons for this. Just twenty years ago, both China and India relied on providing Western scientists with access to raw biomaterial and clinical or laboratory labour in exchange for being part of global scientific knowledge production (Fischer, 2018; Jayaraman, 2005a; Salter 2008: 156). Yet a one-way diffusion of science and its governing practices has always been a myth and non-Western societies have always interacted with and reshaped various forms of knowledge from the West (Palladino and Worboys, 1993: 99–102). This is especially true today as China and India have become magnets for international collaboration and the very sites of global knowledge production; respectively they are the first and third largest countries in terms of research outputs (NSF, 2020).

In this book we examine critical events that have shaped the 'national habitus' of China's and India's scientific development in the first two decades of the twenty-first century. By delineating the evolution of the 'national habitus' in both countries, we reject a reductionist approach of a 'lifeless Asia' (Said, 1978: 154), which is fixated on the classic top-down chain of command and tends to treat divergence as rooted in an imagined heritage of 'cultural differences' (Zhang, J. Y., 2021). Instead, we draw attention to the *contemporaneous* nature of how the life sciences are developed in both countries and the layers of *agencies* activated by new and emerging social relations in the framing and addressing of unprecedented socio-ethical contentions both domestically and transnationally. Examining the chain of critical events, which includes both breakthroughs and controversies, provides a particularly fruitful avenue to make visible the *realpolitik* faced by subaltern communities in a global age. By subaltern, we do not mean these communities are necessarily social-economically 'oppressed', but we refer to social actors' perceived position of their unequal relations of epistemic power. More important, as the book demonstrates, scientists and entrepreneurs are exploiting platforms such as the India-Brazil and South Africa Forum, China's Belt and Road Initiative and International Association of Neurorestoratology to explore and establish future-oriented governing architectures in new subfields. Epistemic disobedience is not limited to the reactions of anti-(Western) authorities but is a reckoning of the capacity to act 'out of rank' without the involvement of traditional global elites.

For philosopher Hannah Arendt, science is necessarily a political project for it can only achieve meaning when we are able to 'talk with and make sense to each other' and to ourselves (Arendt, 1958 [1998]: 4). She warned that the inability for us to understand the implications of technological innovation may lead us to be enslaved by our own creations. More than six decades later, the human condition is further complicated by the fact that we are struggling to make sense of each other's scientific ambition (if not intention) and we are entrapped in our own tunnel vision of the epistemic goods and evils. Science is and must be in the service of the societies of which it is part. But as the world becomes ever more cosmopolitan, given contemporary science's almost instant global outreach and increasingly diversified social aspiration, how do we maintain our political capacity to think and speak about science collectively in a fragmented world? How can we effectively proceed in solidarity but without imposed uniformity? How do we uphold both diversity and responsibility? Obviously we do not have all the answers. But a decolonised understanding of the whys and hows of India's and China's epistemic disobedience in contemporary life sciences is the best way to start the conversation.

Acknowledgements

This book draws on findings from a series of research projects the authors have led or contributed to. We would like to thank the Wellcome Trust, the Economic and Social Research Council, British Council and the Ministry of Science and Technology in India for their essential financial support.

We want to express our deep gratitude towards Nikolas Rose for his consistent support, guidance and for being an inspiration over the years.

We give our thanks to the hundreds of people who have helped us during our various fieldwork and to everyone who provided constructive critique of our work. We are grateful to have wonderful colleagues at the School of Social Policy, Sociology and Social Research at the University of Kent, and the Department of Science, Technology, Engineering and Public Policy at University College London. We also want to thank our friends and former colleagues at the School of Global Affairs, King's College London.

In addition, Joy would like to thank: Tracey Brown at Sense about Science and Paul Manners at the National Coordinating Centre for Public Engagement for their friendship and valuable insights; Andrew Allen and Jens Hein at the Royal Society for stimulating conversations and generous support; and finally Michael and the paws for making every place feel like home.

Saheli would like would like to thank: Brian Salter at King's College London and Jahnavi Phalkey, Founding Director of Science Gallery Bengaluru for their mentorship and generous support over the years; Raghavendra Gadagkar and Uday Balakrishnan at the Indian Institute of Science for their insights and thought-provoking conversations; Yinhua Zhou for his unstinting help; and finally Mom, Dad and Richard for bearing with her during her research.

Finally, we are indebted to Senior Commissioning Editor Thomas Dark at Manchester University Press, Inscriptions series editors Des Fitzgerald and Amy Hinterberger, Assistant Editor Lucy Burns and our anonymous reviewers for their input and help.

1

The global science race and the decolonial imperative for governance

Science, as a systematic and methodical social production of knowledge, is ultimately about power. Governing science is about the means and ends of steering that power. Increasing global competitiveness in science, especially from the audacious steps taken by Global South countries such as China and India, has interrupted the pace of how global governance is negotiated and delivered.

Our discussion begins with the unpacking of a scandal, not (just) because China and India are often considered as the hotbeds for fraudsters and rogue scientists – bear in mind that they did not become science powerhouses through unruliness or fabrication of data – but because this scandal both reaffirms and unsettles such conventional impressions, just as it both reaffirms and unsettles the power dynamic that we've relied on to advance global science. More importantly, it is a perfect example which shows the multi-layered ambiguities and contradictions in the *realpolitik* of the Global South's drive for influence in frontier research and how power struggles are enmeshed with subaltern anxieties: to modernise, to be recognised and to be respected despite their positional disadvantages in the global epistemic hegemony of science (Santos, 2002; Spivak, 1988). As philosopher Kwame Anthony Appiah (2006: 77–8) insightfully put it, 'warring parties are seldom at odds because they have clashing conceptions of "the good." On the contrary, conflict arises most often when two peoples have identified the same thing as good.'

The perfect scandal

When Joy's long-haul flight landed in Hong Kong on 26 November 2018, she was taken aback by several dozen messages that simultaneously burst onto her phone: Chinese biophysicist Jiankui He, based at the Southern University of Science and Technology in Shenzhen, announced the birth of the world's first twin girls with edited genomes. The research wasn't published in peer-reviewed journals and his university in Shenzhen was not aware of the experiment (Cyranoski and Ledford, 2018). But through an interview with the Associated Press, He claimed that he had successfully engineered mutations into human embryos by applying the latest CRISPR gene editing so as to keep HIV-positive parents from infecting their children (Marchione, 2018). Both the news itself and the timing of it were shocking. The reason Joy was in Hong Kong was to speak at the Second International Summit on Human Genome Editing (hereafter the Summit), scheduled to start the next morning. Co-convened by the US National Academy of Sciences and US National Academy of Medicine, the Royal Society of the UK, and the Academy of Sciences of Hong Kong, a key agenda of the Summit was to review and discuss how to proceed with heritable genome editing, an area of research put on hold by the previous global summit in 2015 until further consensus was reached on related safety, efficacy and ethical issues (National Academies of Sciences, Engineering, and Medicine, 2015). With his Associated Press interview and the release of a pre-recorded promotion video from his lab, Jiankui He, who was also a speaker at the Summit, seemed to have completely subverted the regulatory pathway designed by a circle of scientific elites, and, without wider consultation, pushed the world into a new era in which science could rewrite the gene pool of future generations by altering the human germ line (Cyranoski, 2019). The initial response in the West was mixed. While the majority considered He a 'rogue' and his experiment 'not morally or ethically defensible', at least one high profile scientist in the US defended He's attempt as 'justifiable' for combating serious public health threats (Marchione, 2018).

That evening, Joy had dinner with Chinese bioethicist Xiaomei Zhai, a member of the Summit's organising committee. Joy first met Zhai in 2006 in London. At the time, 'bioethicists' as a profession was a much marginalised player in China's science polity. Zhai used to half-jokingly describe her job as 'making a noise' at regulatory meetings so as to get the idea of ethical governance noticed among Chinese decision makers (Zhang, 2012a: 169). However, more than a decade later, there has been some noticeable shifts in China's science politics. That dinner, for example, was interrupted several times by her mobile, as ministries and leading scientific institutions in China scrambled for her views on how to respond to the CRISPR baby case. Zhai has been a veteran in handling such publicity crises for Chinese science, which often involves the tricky business of treading a fine line between upholding professional standards and defending China's regulations. But for Zhai, He's case was easy: the Chinese government had banned reproductive experiments on human embryos, thus his experiment was illegal. Campaigning for the rule of law in the Chinese scientific community has been a key agenda of Chinese bioethicists (see Chapter 3). Regardless of what contextual reasons He might have had to conduct his experiment, it was unjustifiable in Zhai's view.

The focus on the rule of law was echoed by the Chinese scientific community. Within 24 hours, 122 Chinese scientists signed a joint statement condemning He's research as 'a huge blow to the international reputation and the development of Chinese science, especially in the field of biomedical research' (ScienceNet, 2018). Moreover, the statement stressed that damaging China's reputation was 'extremely unfair to the large majority of diligent and conscientious scientists in China who are pursuing research and innovation while strictly adhering to ethical limits' (ScienceNet, 2018). In the eyes of Chinese life scientists, precisely because cutting-edge research was always at the borderline of ethics and legality, adhering to rules, where they existed, was all the more essential to reverse the impression of a reckless 'Wild East' (Dennis, 2002; Song, 2017).

Towards the end of the joint statement, Chinese scientists urged their government to tighten regulations in this area as 'the Pandora's box has already opened, but we may still have a chance to close it before it is irreversible' (ScienceNet, 2018). But two days later, while repeatedly stressing to the public that heritable germline editing was 'irresponsible and that He's self-claimed success remain to be verified, the organising committee of the Summit concluded that 'it is time to define a rigorous, responsible translational pathway toward such trials' (Baltimore et al., 2018). By reorienting the goal from seeking 'broad social consensus', as outlined by the first Summit three years earlier, to identifying 'translational pathways', the organising committee, predominantly North American scientific elites, effectively decided to keep the Pandora's box open.

The decision immediately drew scepticism from the bioethics community (Begley, 2018; Dickenson and Darnovsky, 2019; Hurlbut, 2021). It was perceived to be motivated by the technological determinism that drove contemporary science: 'the first person who puts it on paper wins', which was also part of He's logic (Kirksey, 2021). This made criticism from the scientific establishment look like a charade, for their real intention seemed to be to maintain Western authority over who and how science could be conducted rather than to safeguard public welfare. A Chinese biologist told Joy that, while he remained strongly opposed to He's work, he found Western rhetoric about this case ironic: 'It reminds me of the Chinese saying: "the governor can set the fire but the governed cannot light a lamp!"' Such scepticism towards a Western 'double standard' is neither new nor restricted to China. A year later in London, while Saheli was comparing the ethical regulatory gaps of experimental therapies between the West and India, she was struck by the irony that the UK tabloid newspaper the *Daily Mail* had just reported a Ms Reema Sandhu who paid £70,000, sourced partly through GoFundMe, to the private HCA London Bridge Hospital for experimental stem cell therapies. The UK media heralded such case as 'reap[ing] remarkable benefits of medical revolution', while similar practices in India became the evidence of an irresponsible and exploitative research community (Waters, 2019, see Chapter 4). For some Global South scientific practitioners, there seems to be a pattern of a discretionary interpretation of events.

We argue it would be a trivialisation to consider He's case an act of a rogue scientist, or a 'Chinese' scandal. In fact, as subsequent discussions in the West show, the CRISPR baby case underlines a deeper tension between cutting-edge science and its stewardship, which is shared globally. 'He had an awful lot of company to be called a "rogue"', as George Church, a leading CRISPR scientist at Harvard commented (Cohen, 2019a). Jiankui He, who acquired his doctorate from Rice University, had kept in regular contact with his American mentors and formed a 'circle of trust' amongst leading scientists in the US, including a Nobel laureate and social scientists (Cohen, 2019a). In a letter titled 'Did a permissive scientific culture encourage the 'CRISPR babies' experiment?' to *Nature Biotechnology*, ethicist Donna Dickenson and policy advocate Marcy Darnovsky (2019) were not referring to Chinese policies, but to the 2017 US

National Academies of Sciences, Engineering, and Medicine report which concluded that gene-edited children were ultimately permissible to prevent the birth of children with serious diseases. Their piece also referred to the UK's Nuffield Council on Bioethics 2018 report which followed a similar line of reasoning. In fact, to reassure his hospital ethical review board of the legitimacy of his experiment, Jiankui He cited the same 2017 US National Academies report, which provided justification for related laboratory research already underway in China, Sweden and the UK (Regalado, 2018). According to biomedical historian J. Benjamin Hurlbut who had several long conversations with He, he had 'a very earnest desire to move the science forward' and 'want[ed] his effort to count for good' (Begley, 2018). He's fault lies not in ignoring Western debates on this issue, but in 'listening too intently' to views circulated in 'the inner spaces of science' (Hurlbut, 2021). This made He believe germline editing was inevitable, and all it needed was 'a person to break the glass' (He in Hurlbut, 2021). When He saw a survey that showed the support of genome editing to prevent HIV infection was 73% among the Chinese public, and 95% among those with HIV, he was convinced that, despite being controversial, 'ethics are on [his] side of history' (Begley, 2018; Normile, 2018).

While He's violation of Chinese legislation is indisputable, to what extent his experiment was 'ethically' problematic, and for what reasons, remains a global debate. Drawing on the history of modern biomedical advancement, some may even argue that a question mark could be put over whether, in the future, the CRISPR baby will be perceived as a Jesse Gelsinger or as a Louise Brown event. Jesse Gelsinger was the first publicly identified person who died from a gene therapy trial in 1999, while Louise Brown was the world's first test-tube baby brought to the world in 1978 by Robert Edwards and Patrick Steptoe. But when Edwards and Steptoe carried out their trial, it was also seen as 'irresponsible human experimentation' with limited risk assessment and a lack of meaningful informed consent (Meagher et al., 2020: 331). Despite being sentenced to three years in prison, Jiankui He definitely saw himself as being in Edwards and Steptoe's footsteps (BBC, 2019). Expert examination of He's data concluded that the Cas9 enzyme he had worked on had cut the genome at the correct target site, but instead of producing a full 32-base-pair deletion, one of the twins had a 15-base-pair deletion, while the other had a four-base-pair deletion (Zimmer, 2018). In other words, He only achieved part of his desired mutations in the genes. As such, both the benefits and risks to the twin girls remain unknown.

At the time of writing, this was the latest major scandal in the global race of taking advantage of CRISPR gene editing technology. But what was the scandal? Was it He's 'lack of transparency and the seemingly cavalier nature in which he embarked on such a landmark, and potentially risky, project' (Cyranoski, 2018a)? Was it the Summit organising committee's cunning reframing of regulatory priorities of germline editing research? Or was it the dominant global research culture 'that puts a premium on provocative research, celebrity, national scientific competitiveness, and firsts' (Begley, 2018)? Depending on one's perspective the answer could vary. Or it could be all of them.

The real shock was that it was an obscure researcher from a university which, at the time, was only seven years old that brought forward the global timeline of both scientific and regulatory discussions in this area. A field that has been dominated by Western discourses, with well-established pecking orders and differentiated expertise among research institutions, needs to get used to responding and anticipating the disruptions and derailment from these capable newcomers.

The CRISPR case is also instructive in helping to contextualise the attempted and potential subversiveness of these newcomers. Here we make several points. First, while the challenges from an expanding landscape of global science are real, it is important to highlight that disputes over research agendas are *not simply due to a West vs East divide*. In fact, as the CRISPR case showed, Jiankui He's research was condemned by both his international and domestic peers, in line with the scepticism towards the Summit's call to further clinical research of germline gene-editing. The kernel of discontent is more often on whose opinion is given what weight, on what grounds, and/or for what purpose.

Secondly, and relatedly for latecomers in global science, such as practitioners in China and India, there is an acute awareness of a power imbalance in the global epistemic hierarchy and their image of the 'Third World scientists' (Prasad, 2005). As later chapters demonstrate in detail, for these previously marginalised actors, developing research capacities, both at the individual and the collective level, is necessarily a subversion of this political dominance. The point of this subversion is not antagonism but *a struggle for recognition*, as an equal and respected partner in the global ordering of scientific practice. For example, the short statement from the 122 Chinese scientists criticising He highlighted the value of the rule of law and how the majority of their peers 'strictly adher[e] to ethical limits' (ScienceNet, 2018). Conformity was not an issue. In fact they were eager to demonstrate they were cooperative and trusted players. Yet the lingering shadow, as the Chinese biologist's 'fire/lamp' analogy entails, was to what extent they could have a sense of ownership and belonging to the global order of science.

There are, of course, historical reasons for prejudice and stigmatisation in both the Global North and the Global South. But it's useful to bear in mind that for both China and India, their scientific agenda is forward-looking, rather than trapped in their past. This is why we want to underline a third point: Jiankui He's international connectedness reminds us of the *contemporaneity* of Global South actors. He was not guided by traditional Chinese thought or dreams of national rejuvenation but was reacting to the stigmatisation of HIV in China and to the trend in Western academic debates to relax germline research. China's regulatory shortfalls certainly were a big enabler. Local particularities still exist and are still very important. But such particularities are not trapped in each locality's ancient history; rather, they are developed through reciprocal interaction with the rest of the world.

Finally, the CRISPR baby's regulatory provocation did not end with Jiankui He. In the months that followed his announcement, Russian geneticist Denis Rebrikov revealed that he had long planned to create gene-edited babies to treat inherited deafness and was in the process of seeking ethical review (Cohen, 2019b). This was not as disturbing as the news from Austin Texas, where a 29-year-old programmer Bryan Bishop and a former biotech company lab scientist Max Berry started their human germline engineering company funded by Bitcoin and already had one parent-couple customer lined up (Regalado, 2019). This echoes a general worry at the International Summit in Hong Kong when Jiankui He presented his CRISPR baby case. That is, regardless of what policy path China or the UK, or US authorities respectively decide to take, it can hardly address the worry that somewhere, someone will do it again. A very thorny policy problem shared worldwide today is that while national policies and institutional censure may keep professional scientists with public funding in check, the ability to keep in line the fast growing constellation of experiments that are outside of the conventional scientific system, and thus outside conventional political accountability, is very limited.

Following the Summit, the three presidents of US National Academy of Medicine, National Academy of Sciences and Chinese Academy of Sciences co-authored an editorial in *Science* and referred to He's experiment as a 'wake-up call' for science governance. For solutions, they pointed to the Asilomar Conference on Recombinant DNA in 1975 as a 'model' to follow (Dzau, McNutt and Bai, 2018). The Asilomar Conference was a revolutionary act where a group of 140 professionals, mostly molecular biologists but also lay people and physicians, met in California and publicly debated safety and efficacy concerns before reaching consensus on a set of guidelines that became foundational for government policy. But the conference ascribed value only to narrowly defined expert opinions and its success was driven by the view that technoscience was the driver of social change (Hurlbut, 2015). Furthermore, public engagement of science has long since exceeded the depth and complexity of the Asilomar Conference. More importantly, over 40 years on, global science is no longer restricted to research institutions or to Western societies with similar socio-economic backgrounds. Thus, there is a need for more radical rethinking of science governance at the global scale.

The decolonial imperative of science governance

Decoloniality is an epistemic project. Whereas the concept of 'coloniality' points to a vertical global power structure and the term 'post-colonial' refers to the understanding and dissection of that power in various aspects of socio-political life, decoloniality has a more specific focus on the anti-hegenomic endeavours to counter epistemic power imbalances (Harding, 2016, 2019; Mignolo, 2011; Mignolo and Walsh, 2018: 18–24; Quijano, 2000). In the field of science and technology studies, it challenges the dominance of a

Western or Northern rationality as 'the one and only desirable and universal' framework to appreciate and guide research and innovations (Harding, 2019: 49).

Decolonial theorisation is not just about the oppressed, but underlines the value and necessity of comprehending how different communities are living in 'different "reals"' (Harding, 2019: 61). So it is not about replacing or competing with Western dominance, nor is it simply about tolerating different worlds. It transcends an old 'either/or' logic, in which science is either regarded as representing universal truth or as a social product of irreconcilable localism. Rather, decolonial thinking is about taking the evolving and conditional nature of science seriously, and about learning to make use of the varieties of ways to understand our relations with the social and natural world. A decolonial lens is not limited to previously colonised societies, but, as Anderson and Adams (2008: 184) have rightly argued, should include 'the unequal and messy translations and transactions that take place between different cultures and social position ... even within Western Europe and Northern America.'

Modern (Western) science is also built on colonies. At one level, this refers to how a Eurocentric view of science has been a politically and economically powerful tool in regulating the development of non-Western societies (Hamilton, Subramaniam and Willey, 2017). But at another level, we stress a more fundamental meaning of the word 'colony'. Scientific colonies can be seen as institutionalised power structures that govern a group of individuals with similar interests or are committed to a certain type of behaviour. It is a social space and interactive order that does not need to be physically bounded to a specific geography, but can be a transnational assemblage of 'material, collective, and discursive relationships' (Collier and Ong, 2005: 4–12; Sassen, 2006).

The decolonial imperative of science governance that we argue for is a call for a radical rethinking of how we govern, as a collective, and in recognition of the heterogeneous, contingent, dialogical and situated nature of contemporary science practice, especially in the field of emerging life sciences. The decolonial imperative urges 'thinking *from and with*' and not simply '*about*' people, subjects, places and their relations (Walsh, 2018: 25–32). We achieve this through the analysis of China's and India's 'national habitus' and a focus on critical events. In this section, we want to first list three key directions relevant to this book on how science is spinning out of control from conventional colonies of authorities.

Bottom-up brokerage

To some extent, every research project is a 'bottom-up' initiative. An idea does not parachute from the sky but has to come from someone or somewhere. But until recent decades, the organisation, legitimisation and validation of modern science have mainly been the remit of formal institutions. When Kuhn (1962 [1996]: 76) analysed paradigm shifts in science, he noted 'invention of alternates is just what scientists seldom undertake'. The reason for this seemingly 'conservative' attitude was a practical one. 'So long as the tools of paradigm supplies continue to prove capable of solving the problems it defines, science moves fastest and penetrates most deeply through confident employment of those tools' (Kuhn, 1962 [1996]: 76). Yet, a few decades later, opposite to Kuhn's description, there seemed to be a growing enthusiasm in exploring alternative reasoning and in applying plural interpretations to a given set of data. This 'greater versatility' (Beck, 1992: 167) observed in contemporary scientific endeavours is not simply due to factual ambiguity or inadequate information but has more to do with a social mentality in that practitioners are aware of the limit of their (and other's) knowledge. To paraphrase Dr Jonathan Kimmelman's remark at the release of the 2016 ISSCR Guidelines for Stem Cell Research and Clinical Translation (EuroStemCell, 2016), science moves quickly, and scientists recognise guidelines are not doctrines but 'living documents', which are subject to ongoing review, interpretation, and revision as new evidence emerges. With this mindset, scientists do not blindly submit to a discourse, but 'shop for' or actively cultivate socio-political environments that can accommodate their research agenda (Russo, 2005).

This tendency towards 'shopping' was further made possible with the rise of the East. In her study of India and Japan, Sleeboom-Faulkner (2019: 372) found that when individual scientific practitioners acted as the intermediaries in organising international collaborations, they were also acting as 'regulatory brokers' who transform their geographic knowledge about different regulatory regimes into (scientific and financially) profitable 'regulatory capital'.

Such 'brokerage' was not a solely reactive strategisation of existing regulations but an independent source of influence to the 'active creation or reform' of national and international norms (Sleeboom-Faulkner, 2019: 359; Sleeboom-Faulkner and Patra, 2011).

Networks may empower individual actors but this does not mean equal empowerment of all partners or that everyone is empowered at the same time. Networks are not egalitarian, rather, some nodes naturally have more leveraging power than others, and some would benefit more than others (Sassen, 1991). The density and scope of collaboration remains a reflection of global status. But networks are transterritorial places where bottom-up brokerage of resources, be it material, technical or political, can be transformed into new forms of sustained influence.

De-territoriality of science and its governance

The transdisciplinary and transnational nature of contemporary science has been so normalised that it has almost become redundant to point it out. This is especially true in the life sciences, where the streamlining of technical specialties, the need for onsite data collection and off-site analysis have demanded cells, animals, chemicals, patients, and scientific professionals to travel and interact across borders (Fearnley, 2020; Ong, 2016; Sleeboom-Faulkner, 2019; Song, 2017; Zhang, 2010). Stem cell therapy research is a prime example where scientific practices become an assemblage of their own and do not neatly fit in with previous disciplinary or judicial categorisations (Datta, 2018; Song, 2017; Tiwari and Raman, 2014). Diversification of funding challenges the dominance and necessity of institutional support. The diversity and multiplicity of projects stemming from the bottom up also leads to a disconnect between scientific practice at the bench and research planning at government level (Prasad, 2005; Regal, 2018; Salter, Zhou and Datta, 2016; Sleeboom-Faulkner and Patra, 2011).

What makes decolonialisation an imperative for thinking about governance is that such de-territoriality described above decouples 'identity space' from 'decision space', which renders conventional geopolitical theorisations of the world invalid (Maier, 2000). That is, the geographic, institutional or disciplinary anchoring of a person or a project may no longer be congruent with the turf that provides funding or assurances of legitimisation. As such, de-territoralisation impinges and transforms how responsible research is identified, tested and enforced. Relatedly, there is also a disaggregation of projects from networks where an idea or practice may spin-off from the network that gave birth to it and migrate to other socio-political niches to establish relevance and impact (Rankin, 2017). De-territoriality of contemporary science does not mean territories don't matter, but it highlights a possible incongruence between how we do and how we govern science. Boundaries, along with the power that comes within them, are still important. But they would exert more importance if decision space once again corresponded with identity space, if global governance truly spoke for and not simply to the global.

Cosmopolitanised civic epistemology

It has long been recognised that the boundaries between the natural sciences and the social sciences are not clear cut. The ontology (i.e. 'things') and epistemology (i.e. the way we know and think about 'things') are 'hybrid' (Jasanoff and Simmet, 2017; Latour, 1993). In her seminal work, *Designs on Nature*, Sheila Jasanoff (2005a) powerfully argued that scientific knowledge can only establish its authority when it is articulated, deliberated and valorised in a way that meets entrenched cultural expectations and conforms to public reasoning. This socio-politically grounded 'knowledge-ways' through which citizens 'assess the rationality and robustness of claims that seek to order their lives' is what she defined as civic epistemology (Jasanoff, 2004; 2005a: 249).

However, recognition does not often get translated into action. Even in countries with a strong culture and infrastructure for public engagement, such as the UK, the quality of public dialogue remains prone to be eroded by 'an elite scocio-technical imaginary of "science to the rescue"' (Smallman, 2020: 592). However, social studies in genetically modified organisms and biobanking, for example, have all underlined that in a globalised yet fragmented world, doing science has become less of an act of predestined necessity and more of a socially mediated choice (Levidow and Carr, 1997; Salter and Jones, 2005).

To some extent, the public has grown more mature than the polity that governs science. Policy deliberation both in the Global North and particularly in the Global South remain – explicitly or implicitly – guided by the 'ELSI' framework, a legacy of the Human Genome Project (HGP) in the late 1990s, which assumes that ethical, legal and social issues are not only separate from science, but can be dealt with in their respective epistemic boxes (Balmer et al., 2016). Meanwhile the public increasingly sees both benefits and risks as inextricably bound to a technology under discussion and embraces an imaginary of science only as a force for 'contingent progress' (Smallman, 2018: 668). In addition, the expansion of the civil sphere from the national to the global is not just an inflation in size, but also infers a qualitative leap: civic epistemology has become 'cosmopolitanised' (Zhang, 2012a). This is to say, in addition to the diversity of civic reasoning, 'knowledge-ways' from different communities may also interact, inform and transform each other. It is safe to say that civic epistemology in the global age has become more reflexive and contingent, which, as this book demonstrates, presents both challenges and opportunities for governing responses.

In short, bottom-up brokerage, de-territoriality of science and cosmopolitanised civic epistemology are three key trajectories of contemporary science which necessitate a decolonial reflection on our governing approach. To decolonise, we need to be able to 'de-link' ourselves from a Western and Northerly perspective and acquire the ability to think *from and with* the developing countries (Mignolo and Walsh, 2018). There are various ways to achieve this. One of the approaches, as we propose in relation to China and India, is to comprehend their evolving 'national habitus' of the life sciences through critical events in recent decades.

'I am where I think'

If the world is becoming more diverse, fragmented, intersectional and, more importantly, more fluid, how are we ever going to apprehend agency? Decolonial theorist Walter D. Mignolo (2011: 92–3) suggested turning Descartes' famous dictum around: instead of 'I think therefore I am', we are in a world of 'I am where I do' or 'I am where I think'. It is our evolving outlook in a matrix of power and our ability to afford such outlook that constitute our biographies.

Thus, to unpack rising scientific powers, such as China and India, we need to understand how they situate themselves in the confluence of global exchange, and how such positioning is informed by and informs their ambitions and actions. In this book, we achieve this through elucidating the 'national habitus' of the two countries. To illuminate 'national habitus' is different from making normative (and static) characterisations of something as being 'national' (Yair, 2019). Similar to many other empirical social scientists, we do not believe the two countries behave differently simply due to enduring cultural traits or a lack of 'scientific values' (Prasad, 2005: 465). Nor do we think such views are helpful. Repacking ignorance into normative boxes does not make it disappear, it only creates more obstacles in actually addressing it. Indeed a key value in examining China and India lies in the fact that many of the hopes and worries they have generated are reflected in the emergence of other new powers. But given the intensity and scope of their scientific venture, their global ascent has a magnifying glass effect on these issues.

'National habitus' helps to keep in sight the dialogical relations between a changing China/India, and the world's science politics that they have changed. Building on the work of Bourdieu and Wacquant (1989; 1992), we define national habitus as durable and transposable dispositions of scientific systems within a national jurisdiction. They are a result of cumulative exposure to a set of social relations and conditions and inform the individual and collective schemata of perception, appreciation and action. This also gives insight on what Michael Fischer calls the 'peopling of technology' (Fischer, 2013a; 2013b). Networks and modes of knowing are not givens but are outcomes of the biographies and choreographies of scientists who actively sense, select and respond to technical possibilities.

More specifically, we illuminate national habitus through the examination of critical events that have shaped the course of China's and India's domestic discussions on science development since the late 1990s. Scientific breakthroughs, as well as controversies such as the genetically modified crops, embryo research, experimental stem cell therapies and COVID vaccines, present key moments where national and international discourses come into conflict and eventually come to a compromise. Examining critical events provides a fruitful avenue of deciphering collective thought process in countries like China and India (Das, 1996). The deliberations over the socio-ethical governance of the life sciences in both countries are much less structured and institutionalised than in many Western countries. This contextual characteristic is widely known and often criticised. But its cause and impact are not well understood. Our discussion goes beyond giving simple normative judgements on what science deliberation should look like. Instead we examine how norms and debates are 'pushed' by global scepticism or outcries over novel practices in these two countries. Issues such as scientific uncertainty, research legitimacy and (national and international) accountability are repeating themes that underpin the clashes between the 'global' (Western) and domestic views. These are also areas that Chinese and Indian stakeholders have gradually learnt to negotiate through experimenting with different approaches over the years. As such, critical events function as 'contact zones' in which 'disparate cultures meet, clash, and grapple with each other … in the presence of asymmetrical relations of power' (Pratt, 1992: 7–8). Thus, national mentalities and collective actions are not necessarily reflected in policy responses to a single technical issue but are exhibited in a chain of events a particular community faces over the decades with evolving socio-political dynamics.

Structure of the book

China and India are seen both as the dragons and elephants in contemporary life sciences. As 'dragons' they both embody a forceful drive to excel, yet as latecomers they are also the 'elephants' that tip the balance and subvert the norms of global science. In this chapter, we've laid out why some of the 'deviance' manifested by China and India are not country-specific, but underlie shared challenges brought on by a growing diversity of ways of doing research outside of conventional institutions. To catch up with this change, there is an urgency to decolonise our way of governing science. To decolonise science, governance cannot be a one-sided task but requires genuine rethinking from both the Global North and the Global South.

Chapter 2 sheds light on the subaltern anxieties shared by China and India in order to help untie a Gordian knot of mutual scepticism between the West and the new powers in the East. Seen from the West, China and India often occupy a 'geography of blame' where their aggressive scientific agendas provide fertile ground for fraudsters and mavericks. Western observers thus argue Chinese and Indian scientific communities need to first prove themselves as trusted players in order to win respect. Yet, in the eyes of many scientific practitioners in China and India, they are unfairly condemned to a 'geography of victimisation' due to a long-standing epistemic injustice. They argue that the West needs to acquire a fair attitude first so as to appreciate the actual scientific contribution from the two countries. This Gordian knot leads us directly to a thorny question: can there be epistemic inequality *within* the contemporary life sciences? More importantly, how would this inequality shape our actions, and inactions? We unpack these questions by elucidating how China and India position themselves in the twin process of modernisation and globalisation. This provides an insight into why mutual scepticism persists and how it can be overcome. The empirical overview on the two countries' developmental trajectory also contextualises discussions for subsequent chapters.

Chapters 3 and 4 respectively delineate the national habitus of China and India through critical events in the past few decades. It is useful to note that our discussion is not intended to be a comparative study of the two countries, but we see them as insightful complementary cases. Domestically, while both countries are confronted with domestic disparities, chronic deficiencies in ethical oversight, and increasing entrepreneurship in science, they adopt different strategies in increasing their scientific capacities. At the international level, even though the two countries have different strengths and political systems, their deviance from Western norms points to similar gaps in how we approach science governance.

Chapter 3 argues that various conflicts and contradictions in China's rise in the life sciences are best understood as a 'struggle for recognition' both domestically and globally (Honneth, 1996). We first set out the basic governing structure and major policy initiatives in China. But such structures should not be seen as static. In fact, through examination of critical events such as China's joining of the HGP, hybrid embryo research, the Golden Rice controversy and the COVID pandemic, we demonstrate that even in an authoritarian country, the national habitus of science is constantly challenged and reshaped by bottom-up initiative from scientists, bioethicists and the general public. More importantly, we highlight the de-territorised nature of these initiatives. We debunk the erroneous impression that researchers in China are passive 'state scientists'. Rather, similar to bioethicists, they actively draw on resources transnationally to establish their professional autonomy and authority within and outside of China. In cases such as the International Association of Neurorestoration, the rise of Chinese-led but trans-nationally organised science has formulated alternative ways of validating knowledge *within* contemporary Western science. This point is echoed in Chapter 5. Some of the governing problems the Chinese government faces echo what is shared worldwide, such as the public both increasingly relying on science and becoming more sceptical to (state) scientific authorities (Gluckman and Wilsdon, 2016). Others are new, such as the emergence of bioethicists as a profession. Despite the establishment of the National Science and Technology Ethics Committee in late 2020 in response to the CRISPR baby scandal, China has yet to have an anticipatory form of governance that incorporates social science and the public's voice. Key themes of this chapter are further developed in our examination on India.

Chapter 4 takes up India's ambivalent relations with its pursuit of *Atmanirbhar Bharat*, translated as self-reliant or self-sufficient India. The ambivalence is rooted in the seemingly paradoxical fact that global openness, or rather, *global-reliance* is the route for achieving *self-sufficiency*. Through examining critical events such as the Bt brinjal controversies, the global outcry of its experimental stem cell therapies, and the promulgation of its latest fifth national science, technology and innovation policy, this chapter also focuses on shifting social relations. More specifically, it juxtaposes scientific elites with the masses, as the *haves* and *have-nots*, which refers not only to their differentiated position in having technological expertise, but also in having a voice in steering science. This has led to the phenomenon of what we call the 'bureaucratic amplification of credibilities', which has been counterproductive in gaining public confidence. The shaping of national habitus of Indian science has always been 'haunted' by Western discourses. But entering the twenty-first century, India, similar to Chinese cases in Chapter 3, seems to haunt Western-dominant regulatory rationales with a new question: 'who *could* do science'.

Chapter 5 examines the co-dependence and rivalry between China and India and their implications for global governance. We invite and enable the readers to think *from and with* the two countries by pointing to the often ignored leftist science populism that underlines Global South societies' management of the dual task of modernisation and globalisation. This helps to identify the latent effects in the two countries' selective global outreach and to understand their limits in leading South–South collaboration. The COVID vaccine diplomacy exhibits the two countries' latest struggle to gain a better position in the global epistemic hierarchy. Whereas China's vaccine diplomacy can be summarised as 'contrast, collaborate and calumniate', India adopted an approach that resembled 'contest, convert and control'. Yet they both experienced some setbacks due to a deficiency in soft power, which is necessary to bring quality change in how science is applied and evaluated.

Finally, Chapter 6 brings together the themes and cases visited in the book and asks what a de-colonised global governance may look like. The book ends with an invitation to ponder the question 'what global science will have been'. This future anterior framing was first proposed by the feminist scholar Tani Barlow (2004). This linguistic construct draws attention to the fact that the anticipated future is embedded in the present (or that a present scenario was embedded in the past). More than at any time in world history, the sciences, especially the life sciences, are shaped by the confluence of private pursuits, national ambition and transnational assemblages. Thus, to ask the question 'what global science will have been' is to draw attention to current power struggles and resource imbalances that both stimulate and confine emerging sciences.

2

Unpacking the subaltern anxiety through modernisation and globalisation

A Gordian knot of mutual scepticism

There is no shortage of research on China's and India's life science governance, nor on the policies and institutional reforms they could adopt. Seen from a Western-centric view, it is quite evident what needs to be done for the two countries to be better global players. Indeed, decision makers in both countries have, to a great extent, espoused this view as well: In the life sciences, both China and India had a policy spree in the years around 2010 (Mani, 2013; Zhang, 2017). At least on paper, the two countries' regulatory structure and policy initiatives are very similar to what one would find in leading Western countries, such as the UK and the US (Kaiser and Normal, 2015). Yet, in practice, both countries have 'jurisdictional ambiguity', with policies and institutional rules often resembling 'ideological statements of intent' rather than actual intervention in research conduct (Prasad, 2005: 474; Regal, 2018; Tiwari and Raman, 2014: 416). As Aihwa Ong argued, this is characteristic of 'post-developmental' societies, which favour differentiated implementation of policies and the disciplining of its research force so as to maximise potential socio-economic benefits while maintaining control (Ong, 2006: 18–21; Ong, 2012). But tangible socio-economic rewards aside, at least in governing the life sciences, there is another key factor underlying Chinese and Indian researchers' and regulators' ambivalent attitude in fully adopting and enforcing Western standards in practice: a subaltern anxiety to advance and to be recognised. The term 'subaltern' in this book refers to a social actor's perceived position in the global epistemic hegemony of contemporary science. As noted in the Preface, following Gayatri Chakravorty Spivak (1988) and Boaventura de Sousa Santos (2002), we use the term subaltern not simply to describe the socio-politically 'oppressed', but to refer to developing regions' marginalised status and unequal influence in a Western-dominated discourses and practices.

China and India would perhaps also use the dragon/elephant metaphor to describe their experiences on the global stage but for a different reason. They would, quite rightfully, consider themselves as rising 'dragons', for both countries have made tremendous progress. Currently the world's second largest investor in science, China has sustained double-digit annual growth of its R&D spending (Normile, 2020). Such investment has translated into the production of 20% of global science publications with a growth rate twice the world average (McCarthy, 2019). Although India's spending and number of researchers are not comparable to China or Brazil, its publications almost quadrupled between 2000 and 2013 (Padma, 2015; Van Noorden, 2015). In addition to the highly regarded Indian Institutes of Technology, India is home to several world-class centres for science education. It is also one of the world's leading filer of patents, with more than 7% annual increase in 2018 and 2019 (WIPO, 2020).

But they may also feel like 'elephants' in international policy discussions, for despite their sizeable contribution to science, they are not 'seen' or respected as equal partners. Both countries suffer from an image problem and similar to other Asian countries have acquired the reputation of being 'particularly prone to scientific scandal' (Lee and Schrank, 2010: 1231). Whereas China's research and development has been described by leading Western media as a nefarious 'rising red moon' that would 'always be bad at bioethics' (Cheng, 2018; *The Economist*, 2019), India, with its relaxed regulation in the private medical sector, also occupies 'the geography of blame' (Bharadwaj, 2013: 33) for offering contentious experimental therapies, with media portrayal of scientific misconduct being its 'new habit' (Kochhar, 2019).

There is some truth to these negative impressions. The political pressure of catching up with the West and nation-building is often attributed as a key cause for 'cutting corners' with regulatory and scientific due procedures. Corruption and discretionary ethical oversight have been a chronic problem in both countries (Au and da Silva, 2021; Chattopadhyay, 2012; Regal, 2008). In addition, according to a systematic survey on retracted papers among 11,600 peer-reviewed journals between 1978 and 2013, China and India stood out with the largest likelihood of retraction of research involving their scientists as primary contributors (Tang et al., 2020). Perverse incentives, such as offering cash rewards for indexed English language journals, combined with nationalist aspirations, have led to the fabrication of data and the rise of commercial paper-mills (Segal, 2011; Wang, Tang and Li, 2015). But paper factories are neither new nor limited to China and India (Else and Van Noorden, 2021; Fanelli, 2009). In line with findings of other studies, the aforementioned survey identified (in descending order) the US, China, Japan, Germany and India as the top five countries in terms of number of paper retractions (Tang et al., 2020).

Furthermore, many stakeholders (e.g. scientists, clinicians, bioethicist and regulators) in China and India would argue that an underlying global epistemic injustice is just as significant a factor in their 'image problem' as domestic regulatory problems. Epistemic injustice refers to an in-group/out-group discrimination against someone's capacity as a *knower* (Fricker 2007). This could be due to unfair depictions of the knower's credibility or a marginalisation of the interpretative values held by certain group (Fricker 2007: 9–14, 152–61). For example, empirical studies have repeatedly highlighted a common frustration among Global South scientists that their lab results are often discriminated against by international journals (Bharadwaj, 2014; Zhang, 2012a; see also Traweek, 1996). The exasperated claim 'why do Westerners see it but not believe it?' from Chinese neurorestoratology scientists, discussed in Chapter 3, perhaps best captures this frustration. This kind of 'can't be good enough' prejudicial bias aside, welfare priorities and wider developmental values pertinent to the Global South often carry limited weight in global discussions, such as the Bt brinjals case discussed in Chapter 4. In addition, Chinese and Indian science practitioners would also point to the fact that while both countries are destinations for medical tourism, regulatory grey areas are also exploited in countries such as Australia, Germany and the US and many countries are complicit in encouraging medical tourism towards the East (Einsiedel and Adamson, 2012; Martin, 2014; McMahon, 2014; Petersen, Tanner and Munsie, 2015).

Discussion on the solution to the two countries' image problem easily gets entangled into a Gordian knot: Western observers argue

Chinese and Indian scientific communities need to prove themselves as trusted players first and then they would naturally win respect, while Chinese and Indian stakeholders argue that the West needs to acquire a fair attitude first and then they would naturally see the value and ingenuities of the Global South in contemporary science. While Western views and expectations are widely understood, Chinese and Indian discontent are much less discussed.

To untie this Gordian knot, we argue it is important to understand the role of life sciences in the dual processes of modernisation and globalisation. We examine these themes in the next two sections. This gives insight into a particular subaltern anxiety shared by both countries that simultaneously gives rise to their aspirations and vexations. The discussion in this chapter effectively provides essential historical background which leads to our investigation in later chapters.

Life sciences and modernisation

Modernisation for both China and India was an exogenous concept in the wake of Western colonial occupation. Gearing to Western medical views was not only seen as a foundation for their way to modernity, but was also seen by imperial powers as an 'ideology of colonial healing' in civilising Asian societies (Anderson, 2002; Comaroff and Comaroff, 1992: 222). As such, modern (i.e. Western) science, especially the life sciences, has been 'co-constitutive' with colonisation for it perpetuates a particular worldview (Hamilton, Subramaniam and Willey, 2017: 615–16; Rogaski, 2014).

For India, Western imposition of ethical norms on indigenous medical practices began in the mid-1830s (Harrison, 1994). Yet, from the start, the elites' 'modern' clinical norms, chiefly brought in by British medics were at odds with the practicalities of nationwide healthcare provision, and exacerbated inequalities in the access to affordable healthcare (Ramasubban, 2007). Nevertheless the 'cultural authority' of Western biomedicine was quickly embraced, especially among the upper class (Khan, 2012: 69; Prakash, 1999). 'Professionalising' indigenous knowledge and healing practices to appropriate Western standards, championed by Indian elites, was embedded in modern state building and further reinforced 'deep-rooted' domestic hierarchies (Arnold, 1993; Bala, 2012: 73). After independence, the Gandhian ideology of the *'sarvodaya* model of frugality and reflexive indigenous development, represented by the spinning wheel, was squarely rejected by Jawaharlal Nehru, India's first Prime Minister (Prakash, 1999). In his address to the Indian Science Congress in 1947, Nehru envisioned modern India to be built through 'the politician and the scientist … work[ing] in close cooperation' (Krishna, 1991: 5). Similar to China, social science studies and the general public played minimal role in policy making (Sharma, 1976).

For China, 'Mr. Science' and 'Mr. Democracy' were first introduced during the May Fourth anti-imperialist movement in 1919 with an urgency to secure national survival. They were meant to be both an ideological and a developmental force. Only one month after the founding of the new People's Republic, the Chinese Academy of Sciences was established in November 1949. Even during its domestic political turmoil in the 1950s and 1960s, China spent 1.75% of its GDP on R&D, higher than Japan and many other economies at the time (Gu 2001: 206). Yet much of the investment was on infrastructure and defence technologies (Gu, 2001). For the Chinese scientific community, the 'spring of science' came in 1978 when the government encouraged original research in all areas in the pursuit of making science a 'production force' in the national economy (Guo, 1978; MOST, 2005; Suttmeier, 1974). Such 'utilitarian' focus on short-term financial gains remain a key character of the government's take on science today (Shen and Williams, 2005).

While both countries struggled with poverty in the first 30 years of their independence, they took different investment paths. India prioritised 'big science', as more than half of the nation's research and development (R&D) spending went into space, defence and atomic energy projects (Krishna, 1991). In the 1980s, funding decisions prioritised Western interests rather than domestic needs (Chandrashekar, 1995). China first prioritised meeting basic subsistence needs and shifting its economy from agricultural to manufacturing (Wang, 2015).

The 1990s was the turning point for both countries. Economically, both intensified their transition from socialist to market economies; politically, biotechnology and bioscience were recognised as critical to both countries' knowledge economies (Mandavilli, 2005; Rajan, 2006; Ratchford and Blanpied, 2008). Arguably, the 1990s was also the time when being seen as 'Third World scientists', with an implicit division of labour between First and Third World science, started to sting (Prasad, 2005). Both China and India boosted their efforts to tackle the brain drain in the early 1990s (Zweig, Fung and Han, 2008). To entice many of its young talents who were lost to PhD or postdoctoral programmes in the West, China set up a series of exchange programmes and established many new university posts (see Chapter 3). In comparison, India's relatively robust and internationalised higher education system meant its brain drain came at a higher cost. That is, a higher portion of Indian researchers chose to 'educate-then-migrate', resulting in a loss of mature practitioners, who were trained in India but then decided to leave (Kapur, 2007). However, India has long seen its diaspora as a 'brain bank', which would eventually help the growth of science and technology programmes in India (Lavakare, 2013: 195). Since its economic liberalisation in 1991, multinational companies have set up large R&D centres (Krishna, 2001). To some extent, India has benefited from a 'brain gain' as many Indians who moved overseas have returned home to continue their work on global problems (Lavakare, 2013: 198). In both China and India, these highly skilled and internationally networked scientists are also highly entrepreneurial. They not only came back to join their home countries' R&D initiatives, but also set up their own private ventures, seeking intellectual as well as socio-economic recognition (Sleeboom-Faulkner and Patra, 2011; Zhang, 2012a). But in contrast to overseas returnees' self-perception of knowledge elites as a new cosmopolitan class, an old Western impression of what they 'should' be appeared to restrain their ambitions. A common perception of North American and European scientific communities at the time was that Chinese and Indian scientists were good 'clinical labour' to deliver research and to meticulously solve a given problem; but they lacked the sophistication to initiate or lead a project (Fischer, 2013a; Jayaraman, 2005a: 483). Chinese geneticists for example, described the West-East collaboration dynamic as such: 'we [in the West] do science, you do the dirty work' (Fischer, 2018: 227). Similar to other Global South countries, the injustice that Chinese and Indian scientists felt in the epistemic hierarchy of modern science was that intellectual credit seemed to be something unfairly 'reserved' for the West (Harris, 2005).

A further stigmatisation came from an emerging (or 'continuing') pattern of the role Chinese and Indian societies played in their increased participation in global research. On the one hand, the very construct of many global rules makes indigenous knowledge less worthy than Western scientific knowledge and limits the scope for many Global South societies to claim recognition (Bhambra, 2014; Ho, 2006). This is most evident on issues of intellectual property rights, for which the novelty of a claim is hard to challenge unless there is published proof or prior use. In the 1990s, thousands of patent applications were filed in the West involving traditional Indian medicine. One well-known case is that of two scientists at the University of Mississippi, who were granted an American patent for turmeric as a wound ointment in 1995. Yet such practices had been used in India for thousands of years. India's Council of Scientific and Industrial Research was able to challenge the patent only because an ancient Sanskrit text for the use was recorded in a 1953 paper in the *Journal of Indian Medical Association* (Kolte, 2020). This patent was thus revoked in 1997. With the specific aim of putting an end to the misappropriation of Indian traditional medicinal knowledge at international patent offices, the Indian government subsequently invested in building the Traditional Knowledge Digital Library, a massive electronic repository of herbal practice This database has provided the ground for India to file pre-grant oppositions at various International Patent Offices, and has forced the withdrawal or amendment of more than 230 patent applications (www.tkdl.res.in).

On the other hand, there was an increasing awareness that while Western medical ethics purports to be a way of balancing the risks and benefits of clinical research and protecting individuals, it also insists that cultural and economical particularities in the Global South were conditions to be corrected rather than something to be taken into account in shaping global research norms (Simpson, 2018; Traweek, 1996). Given their large and relatively under-educated, financially deprived populations, both China and India have been attractive sites for pharmaceutical companies and research institutions to get clinical trials done cheaply and quickly (Lloyd-Roberts, 2012; Nundy and Gulhati, 2005). During the 1990s, a series of exploitative Western medical research conducted in China came to be known as the 'Gene War of the Century' (Guo, 2013; Xiong, 2021). A major scandal concerned Harvard Professor Xiping Xu who led a local Chinese research team and collected tens of thousands of blood samples from illiterate peasants in Anhui province without proper informed consent (Keim, 2003). The research took place in the mid-1990s and was funded by both the US National Institutes for Health and pharmaceutical companies. An investigatory Harvard team went to Anhui in 1999 and concluded nothing improper had taken place. Gwendolyn Zahner, one of the whistle blowers on Xu's team, later expressed her disappointment that the investigators did not look 'beyond [the] paper work' (Lawler, 2002). In 2001, at the Symposium on Bioethics, Biotechnology and Biosecurity held in Hangzhou, Chinese scientists and ethicists had a fierce exchange with Xu on the ethics of gene research (Shou and Li, 2001). The discussion ended with disagreement. In the years to come, Chinese ethicists often referred to this episode of exploitative bioprospecting as emblematic of the 'Wild West', a rebuttal to developed countries' 'Wild East' derision of China's early regulatory vacuum in the life sciences (Zhai et al., 2019). More importantly, the frustration was not just about Western researchers violating ethical principles per se. The fact that they were criticising China's lack of ethical regulations *while* taking advantage of the absence of such protocols angered many (Xiong, 2021).

In short, while in the mid-twentieth century, Western views of what constitutes a good life and how it should be pursued profoundly shaped both China's and India's modernity projects, they have also embedded a power imbalance that is increasingly haunting both countries and their peoples' self-realisation. China and India started as earnest learners and followers of Western science. Yet as their scientific capacity grew, they were shocked to find that Western impressions of them remained frozen in time. Fundamental issues, such as what counts as a valid scientific question or scientific evidence and where such things can be found, remain defined by a Western-dominated discourse, whereas intellectual contributions from the Global South have been chronically overlooked, undervalued or suppressed (Chambers and Gillespie, 2000). Ironically, in the eyes of internationally trained Chinese and Indian knowledge elites, increased global exposure only accentuated this perception of a mismatch, causing the power imbalance to sting even more. Such frustrations remain to the present day, which has led to a blunt protest from Rao Yi to the *New York Times* that life scientists in Asia 'don't want to be guided by Western people' (Tatlow, 2015). However, as the next section demonstrates, the unequal status was not simply a result of Western arrogance. The real restrictive effect of an epistemic hegemony is that it is difficult to think outside of the grid. Chinese and Indian scientific communities are also reinforcing the very epistemic injustice they protest against.

Life sciences and globalisation

Modernisation is a never ending project and globalisation complicates the paradigm in which its success is identified and evaluated. Given the two countries' long history and recent colonial past, it is not difficult to understand China's and India's yearning for global recognition. But strategies to realise this goal are not solely based on domestic particularities. Rather, how scientific programmes should be delivered and legitimised are framed by the dual concerns of being intelligible (if not accountable) to Western-dominated scientific standards while remaining relevant to local socio-economic pressures (Crane, 2010). As least from the perspective of scientists in the Global South, their collective marginalised status translates into an unequal influence in Western-dominated discourses and practices. This intertwining of a sense of long-lost entitlement to international respect *and* a constant anguish over being misunderstood or worse – being alienated from cosmopolitan endeavours – is what we call the subaltern anxiety.

One may well ask: can there be epistemic inequality *within* contemporary life sciences? To date, most of the studies in the decolonisation of knowledge have focused on the competition for equal social legitimacy between modern and traditional (or Western and indigenous) sciences, which are often of different knowledge 'paradigms' (Kuhn, 1962 [1996]). Can there be meaningful 'epistemic disobedience' when researchers are in the same paradigm (Mignolo, 2011: 54)? For example, at least in the areas this book addresses (e.g. stem cells, neuroscience, genetics, immunology and regenerative medicine), almost all scientists, regardless of nationalities, are trained in the 'Western' tradition, if not directly in Western institutions. In addition, let's not forget the British physicist John Ziman's (1978

[1996]: 3) widely cited definition that science is essentially 'a consensus of rational opinion over the widest possible field'.

But let's also not forget that consensus is not a negation of contention, but rather, a particular outcome of contention. As Kuhn (1962 [1996]) elegantly demonstrated, modern science emerged and evolved through a successive and systematic discrimination of facts and their interpretation. But more importantly, a globally heightened consciousness of scientific uncertainties has made the practice and governance of science more contextualised and value-laden (Giddens, 1999). Sociological studies of standards have shown that they are never intrinsically neutral; rather, their 'objective universality and optimality are hard won victories that can be heavily contested by third parties lobbing accusations of bias and politicization' (Timmermans and Epstein, 2010: 74). As China and India march towards the frontier of the life sciences, they are also entering a terrain where natural phenomena have not yet been converted into 'evidence' and our observation and interaction with these phenomena are yet to be fully coordinated and 'disciplined'. For example, in the early days of stem cell research, how to define a 'stem cell line' and what measurements are best applied on what grounds remained open questions. With every emerging field, there is a brief window of scientific discretion where rules and norms are discussed, contested and negotiated. Once established, those rules and norms have a sustained impact on the practice, evaluation and reasoning of science, but also carry significant socio-economic implications. As demonstrated in later chapters, increasingly, with rising research capacity, Chinese and Indian scientists find themselves in such a 'discretional' period with their Western peers. The transnational alliance over the clinical validation as accreditation of neurorestoratological treatments (discussed in Chapter 3), for example, is a case of 'disobedience' as epistemically marginalised scientists no longer want to be a mere followers but co-owners of new norms and standards. Similarly, South–South collaborations established by India and China (discussed in Chapter 5) examine new international space that curates a different ordering of priorities, value interpretation and indeed a different way of positioning their influence.

Pierre Bourdieu's insights on power relations may help to comprehend how epistemic deference and power imbalances have shaped subaltern societies' cognitive framework (Bourdieu, 1969; Go, 2013).

It is useful to be reminded that subaltern is not an identity but a *position* (Spivak, 1999). More specifically, it is a social actor's perceived position of their unequal *relations* of power, their marginalised role in epistemic hegemonies. The position of the subaltern is not a place to stay, but always prompts the urge to transit from the margin towards the centre (Sharp, 2009). There are three interrelated 'thinking tools' to a Bourdieuian theorisation of power and action: habitus, capital and field (Bourdieu and Wacquant, 1989: 50). We've explained the concept of 'national habitus', a system of dispositions (e.g. research capacity, norms, collective memory and ambitions) which structures Chinese or Indian stakeholders' present and future practices. Field is a social space with specific rules and pursuits (Bourdieu, 1969: 77). For example, the global arena of life sciences constitutes a field occupied by different players (e.g. Chinese or Indian scientific communities). The position of these players is structured by their set of capital and their relations to one another. In this field, each player, such as India, strategises from their own habitus on how to deploy their cultural, economic and social resources in exchange for accumulating credibility and influence. More importantly, field position has a structuring effect over epistemic valorisation (Bourdieu, 1990). By the nature of where they are produced, certain types of knowledge and practice may naturally hold more 'symbolic capital' than others.

This helps to explain one ironic fact that while both China and India are weary of Western dominance they are also actively internalising and reinforcing dominant expectations and norms form the West. For example, in examining how Indian clinicians justify human embryonic stem cell therapies, Aditya Bharadwaj pointed out the 'inherently ironic and conflicted' reasoning, which 'is not an ethical stance against consensible scientific hegemony. It actively looks for ways' to 'melt into' it so as to be 'recognized by the scientific community as ethical and practising good science' (Bharadwaj, 2013: 26, 37–8). Internalisation and reinforcing Western norms can be observed in the domestic organisation of research as well. For example, in most of India's science development, an unspoken yet widely understood criterion for having a seat at national committees, such as the 2014 Stem Cell Task Force, was to demonstrate excellence beyond India's borders in international research excellence. It is also the case that those with Western training have a substantive competitive edge over home-bred scientists in getting large grants (Datta Burton, 2018). A similar Westernphilic attitude can be observed in China. As discussed in Chapter 3, having Western endorsement has been instrumental for scientists and bioethicists to legitimise and promote their agendas domestically. Thus, while internationally, China and India are anxious to break free of Western dominance, they are still in a process of learning how to plough ahead without prior guidance. Domestically, both research communities struggle to 'formulate a coherent epistemic view' to inform policy makers on how best to incorporate innovations (Salter et al., 2016: 812). Taking such a subaltern mentality into account, it is perhaps not surprising (although still scandalous) that in the CRISPR baby case discussed previously, Jiankui He cited the 2017 US National Academies report in his ethics application to justify the legitimacy of his research. More specifically, He highlighted that a few months prior, the US National Academies had 'for the first time' sanctioned laboratory research of therapeutic germline gene editing, which would bring hope to cure 'unlimited numbers of' serious genetic diseases (Regalado, 2018). What is stunning is not the implied vision of scientific development as 'a linear program based on deterministic notions of immutable progress', which is a Western import (Rajão, Duque and De, 2014: 768). Rather it was He's confidence – and his confidence in getting ethics committee approval – that it didn't matter that his research was not yet legal in China, for the American academies' view would ultimately override national regulations!

Indeed one should be reminded that the Global South themselves had a role in aggravating their epistemic marginalisation. For until very recently, many developing economies have been preoccupied with 'catching up with the West' with little interest in being 'distracted' by global value debates (Prasad, 2008; Zhang, 2015). In other words, the urge to contest and to conform co-exist. This echoes a central paradox which Spivak (1999: 270–2) has repeatedly pointed out: the way for the subalterns to successfully make themselves heard in the global arena is to adapt to the hegemonic grammar. Instead of subverting or shaking the hegemonic dominance, their move towards the centre may betray their subaltern experience.

Selective contestation and conformity is an essential part of the strategy for subaltern societies to be more globally competitive. It's important to bear in mind that there is competition among Global

South communities as well. As discussed further in Chapter 5, the choice of partnership in science is often enmeshed with wider regional politics (Sleeboom-Faulkner, 2019). Compliance with Western standards could give subaltern societies the advantage for North–South partnerships. For example, upon joining the World Trade Organisation, India decided to steer its pharmaceutical industry away from producing generics and towards the development of new medicines. Since most of the scientific and political leverage remains concentrated in a few multinational pharmaceutical companies, the best way to build the Indian R&D profile and capacity in this field was to collaborate with multinationals. To facilitate such collaborations, in 2005, India harmonised its regulations with international good clinical practice and geared its ethical governance provisions in line with the expectations of Western drug approval authorities (Merz, 2020; Sariola et al., 2015; Sinha, 2008; Yee, 2012). These regulatory changes have enabled India to surpass China as the most favoured Asian country for conducting clinical trials, and have reduced the cost of clinical trials to 20–50% less than what they would have been in developed countries (Sinha, 2008).

Of course, such a weighing of options and trading for advantages is also part of the routine calculation of North American and European partners. As China and India are accumulating more research power, one could argue that, at least in theory, the balance could be tipped the other way in the future. That is, Western scientific communities may compete for China's or India's attention. But at least in the near future, societies in the epistemic Global South remain more vulnerable to the 'global' imaginary, as global recognition and international influence are key validations of their modernisation projects.

Subaltern anxiety and the Gordian knot

Bioethicist Nicholas King (2002: 782) argued that while the imperial goal of global health was 'conversion', whereby the focus was on bringing indigenous medical practices in line with Western mainstream, we should move onto the goal of 'integration' where a decolonised lens helps local practices to be integrated into transnational networks. This is a goal that few in the world today would object to. Yet from a Western perspective, the uncertainty is how to integrate while protecting scientific robustness from bad influence. For countries like China and India, the challenge is their disadvantaged position in vindicating and interposing their views into existing scientific epistemologies dominated by the traditional scientific powers in the Northern hemisphere. For the globally networked knowledge elites from the East, practising science is both the fast ticket to modernity and a constant reminder of their inferior developmental stage. They protest hegemonic dominance but also instrumentalise Western authorities to advance their relative domestic or regional competitiveness. Subaltern anxiety encapsulates the two emerging powers' conflicted sentiments towards their position in the world.

How does understanding subaltern anxieties help us to untie the Gordian knot? According to the Greek legend, the Gordian knot was tied to the yoke of an oxcart by the peasant-turned-king, Gordius, so as to secure the cart, which was an offering to Zeus. To some extent, the Gordian knot of mutual scepticism between the West and the new powers in the East exists because both sides want to secure a space of self-determination and control of the bandwagon called life sciences, a provision to our common future.

There are two versions of how Alexander the Great untied the knot. One version says he simply cut the rope with his sword. A radical rethinking of how science should be governed is perhaps our sword. But where should the rethinking start? A second version of the story says Alexander took out a lynchpin that went through the yoke and by doing so, he managed to loosen the knot just enough for him to eventually untie it. If we could take out the Western-centric thinking, the lynchpin of contemporary science, we might be able to untangle layers of doubts and contradiction. But afterwards, we'll need a new lynchpin for the yoke to work. The following chapters discuss the formation of scientific national habitus in China and India at the confluence of their drive to modernise and to globalise. They also examine how the two countries selectively confirm and contest the power dynamics of global science through collaboration and rivalry. Together, they may illuminate ways to think from and with non-Western societies to decolonise our vision. A truly inclusive way of thinking would surely be a better lynchpin for global science governance.

3

Chinese life sciences' 'struggle for recognition'

There is a short answer and a long answer to explain how China's 'national habitus' enabled its rapid ascent as a global power in the life sciences. The short answer is that it was fuelled by the Chinese Communist Party's goal of 'national rejuvenation', which has been a 'foundational national policy' since 1996 (State Council, China, 1996). As this chapter demonstrates, China's rapid development in research capacity is due to nation-wide (albeit not necessarily 'national') efforts among scientists, bioethicists and the public in establishing their collective credibility and influence in contemporary life sciences. But 'national rejuvenation' also creates a misleading impression that science in China is organised and operated from the top-down. There are significant disparities in research and governance cultures between cities and provinces; in addition, local institutions often have to deal with large discrepancies in the enforcement of regulatory policies and the processes by which to decide what areas of research to invest in. As such, in areas such as regenerative medicine, the government's permissive stand and its increased investment have not always effectively translated into research productivity or social uptake of new technologies (Zhang, 2017). The authoritarian yet fragmented nature of the Chinese science system makes its development trajectory hard to predict.

This is why a long answer is needed to fully comprehend China's success and its implications. One may start by problematising what is to be 'rejuvenated' in the idea of a 'national rejuvenation'. At the geopolitical level, there is an 'image' that needs to be restored. The Chinese characters for 'rejuvenation' (*fuxing*) are the same as for the word 'renaissance'. Yet compared to the European Renaissance in the fifteenth and sixteenth centuries, Chinese rejuvenation in the twentieth and twenty-first centuries is much more outward-looking. It is as much about the revival of domestic cultural vitality as it is about re-establishing China's global reputation as a powerful nation. In the realm of modern science, there is little to be 'rejuvenated' but a lot to be instigated in China. By this we are not simply referring to the famous Needham Question on why modern science did not develop in Chinese civilisation (Needham, 1969: 190). Rather, as this chapter demonstrates, we also point to the 'blank pages' of the regulatory, ethical and public engagement norms that the Chinese scientific system hastily built, under pressure, at the turn of the twenty-first century. It is through such instigations that new forms of social relations emerged in Chinese science. Consequently, one of the main challenges for China is how its particular tradition of politics incorporates and leverages new social relations. This also holds the key to understanding the trajectory of Chinese scientific development.

To borrow a concept from Hegelian philosophy, we argue that the rise of China's life sciences is best understood as a 'struggle for recognition' (Honneth, 1996). The basic Hegelian insight is that mutual recognition is a pre-condition for self-realisation (Honneth, 1996). Thus, an actor's successful integration as a political or ethical subject within a particular community is dependent upon receiving, and conferring, appropriate forms of recognition. This strongly echoes the logic underlying the actions of key actors in Chinese science. That is, striving for development is not simply a search for self-referential betterment. Rather it is a struggle to be recognised both domestically and globally as respected and valued partners of contemporary science. This applies to the Chinese state, research institutions, and individuals. As demonstrated in this chapter, even in an authoritarian state, the status of science or scientists cannot be taken for granted, nor can its national habitus be seen as static. Rather, the development of Chinese life science is a multi-layered process where individual, institutional and state actors simultaneously establish and expand their socio-political space to form new social relations.

This chapter illuminates these struggles for recognition by examining how key relationships within Chinese science have responded to, and been shaped, by critical events. There is a growing body of literature on each individual event examined in this chapter, including works of our own. Thus, we do not wish to simply rehearse a list of mini-case studies. Instead we group them according to the relational tensions they embodied. By doing so, we will effectively review these critical events in chronological order and provide a view of the national habitus of science from different vantage points. More importantly, this will enable us to demonstrate the long-term normative impact these incidents had on Chinese life sciences.

The first section maps out the state's vision of science. It explores the basic governing structure and major policy initiatives China has launched to be globally competitive. For the government, in addition to enhancing economic productivity, 'good science' should contribute to the Party's political glory, rather than stirring up tensions with socio-ethical debates. This led to a 'do first, talk later' research culture (Tatlow, 2015), where the role of regulators was less about avoiding or minimising potential risks than about pragmatically responding to socio-ethical issues in a *post-hoc* manner (Zhang, 2012a; 2017). While state-sponsorship has been crucial for many scientific fields, the government's interest in basic and emerging sciences cannot be taken for granted but needs to be campaigned for.

The second section focuses on scientists' perspectives on scientist–state relations. It examines the Beijing Genomic Institute's (now known as the BGI Group) involvement in the HGP in 1999. This was arguably the first time China's contemporary life sciences became a significant contributory partner in global research. The world's largest genome research organisation, the BGI Group is also commonly (mis)presented as a legacy institute of China's participation of the HGP. We demonstrate that BGI is in fact the product of geneticists' struggle for state recognition of the importance of their research. Furthermore, BGI's success is emblematic of a calculative entrepreneurship that remains essential to science–state relations today. This is partly reflected in the 2018 CRISPR baby scandal.

The third section considers bioethics–state relations. China's scientific rise is also China's bioethical rise in establishing a voice at the international policy table and in forming a professional community of bioethicists. Starting with the hybrid embryo controversy in 2001, this chapter demonstrates that at the turn of the century, the role of bioethicists was one of the 'blank pages' that the Chinese social polity realised it needed to fill. Through the Golden Rice scandal in 2008, and the series of contentions over CRISPR technology between 2015 and 2018, we show how key international disputes were critical moments for bioethicists to capture the government's attention and to fight for a place in (inter)national decision making.

The fourth section discusses how public–science relations in China have been beleaguered by over-politicisation. Similar to how public opinion plays an increasing role in science policy making in the West, in China the expanding middle class is increasingly vocal about having their interests and preferences recognised in local and national agenda setting (Zhang, 2015). A chronic ignorance of public communication and public engagement of science, accompanied by heightened political censorship, has turned a once largely pro-science society into a sceptical one. As social resistance to GM (genetic modification) technologies have significantly restrained the pace of China's scientific advancement in the field, the government and scientific communities started to grasp the importance of attending to civic epistemology.

The fifth section examines science–science relations. Here, we do not refer to the juxtaposition between contemporary and traditional science or between Western and Asian science. Rather we highlight an often ignored power dynamic in global science: the relationship between the international scientific community dominated by the Global North, and the local scientific community of the Global South. A key example is the International Association of Neurorestoration (IANR), which is an international professional association comprised of members from China, India, Iran and Argentina. It grew out of the China Spinal Cord Injury Network (ChinaSCINet) founded around 2004. Once considered as a group of rogue stem cell scientists, ChinaSCINet was also one of a few research models originating in the Global South that was later adopted by countires of the Global North, such as the US and Norway (Qiu, 2009; Zhang, 2022). The development of IANR is a visible and relatively successful example of how researchers who were once considered bit players in global science have gradually acquired both the scientific and political capacities to challenge the epistemic order and boundaries once dominated by the West. Science–science relations can no longer be juxtaposed as only 'Western or Indigenous'; they are also about how actors fight for different research priorities and perspectives to be recognised within contemporary (Western) science.

Finally, the chapter revisits state–science relations through COVID-19. In the eyes of the Chinese government, the crisis brought by SARS-CoV-2 was as much a public health concern as a political one, which went to the heart of 'national rejuvenation'. While we examine China's vaccine diplomacy in Chapter 5, here we underline the impact of government censorship and global media on the openness of Chinese science and their implications for the world. Global media may unwittingly reinforce the nationalist agenda of the Chinese government. At the turn of the twenty-first century, 'China-bashing' may have spurred Chinese science into adopting Western norms and standards. Today the prospect of a hurt but powerful Chinese scientific community reluctant to engage in global dialogue should concern everyone. Building on our discussion in previous sections, we show how China's national habitus is not simply construed by domestic factors but is also formed by a boomerang effect of global interaction.

The state's vision of science: structure, strategy, expectations

It is more accurate to see the Chinese government as the client rather than the sponsor of science. While both roles involve providing substantial financial contribution, a key difference we want to highlight is in the allocation of social responsibility as well as the expected outcomes. In comparison to the role of a sponsor, a client demands the best 'service' money can buy to fulfil its needs (such as helping to achieve the government's economic or political goals) and is more distant from accountability for the (socio-ethical) particularities of a scientific programme. In this section, we demonstrate that both its funding and its administrative strategy reflect this clientele mentality, and have nurtured a linear vision of what 'good science' is (Gu et al., 2009).

The structure of China's scientific system is relatively simple to explain. As shown in Figure 3.1, all branches radiate from the State Council's Steering Committee of Science and Technology. The State Council itself is China's chief administrative authority. It is the Steering Committee's responsibility to coordinate among different ministries and to set overarching goals and objectives in the mid-term and

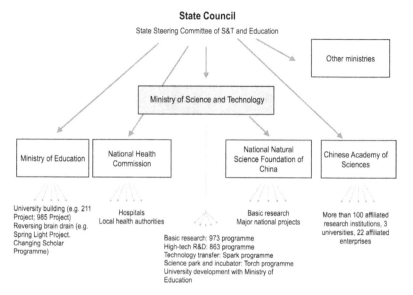

Figure 3.1 The structure of key national administrative offices of China's life sciences

long-term plans on science and technology. All subordinate executive branches, such as the Ministry of Education and the Ministry of Science and Technology (MOST) subsequently translate overarching goals into quantifiable aims or ministerial directives corresponding to their jurisdiction. In short, all regulatory decisions and administrative resources can be traced back to a handful of national-ministry-level organisations. At the same time, state governance reaches local laboratories and clinics through a dispersion of regulations, implemented at different administrative levels. Such structural centralisation was further augmented by China's 2018 initiative of creating a 'science mega-ministry', in which the National Natural Science Foundation of China (NSFC), was consolidated as an administrative sub-branch to MOST with the intention to streamline government procedures and resources (Cyranoski, 2018b: 425). However, currently NSFC remains 'a relatively autonomous institution' in its planning, reviewing, and funding research projects, with a focus on basic research (www.nsfc.gov.cn).

In fact, the actual governance of science is far less centralised or consistent than the structure implies. This is due to a deep-seated *post-hoc* pragmatism in China's science governance (Zhang, 2012a). The government's regulatory attitude towards emerging technologies is often led by a 'do first, talk later' pragmatism, which does not waste energy in precautionary examination of its potential harms and benefits. Rather, regulatory actions are often passive responses to international outcries or scandals that have already taken place (Zhang, 2012a: 32–61). This is in line with China's wider 'no debate' political culture set out by Deng Xiaoping in the early 1990s. As Deng described, to discourage policy debate was 'to save time', for 'debates complicate things and what it abates is time. Nothing can be accomplished' (Gao, 2013: 67). While Deng's original incentive was to quiet domestic scepticism and clear the path for renewed market reform, it also significantly shifted Chinese regulators' orientations of their responsibilities. His idea was that debate does not abate social concerns but only slows down the development that China desperately needed. In the absence of prior scrutiny and discussion, national regulations and guidelines, especially those on emerging science, often appear more as position papers. Vague wordings are often inoperable, and their enforcement can rely on discretionary interpretations of local officials (Rosemann and Sleeboom-Faulkner, 2016; Zhang and Barr, 2013). For the larger part of its modern development, Chinese authorities have relied on regulatory patchworks to govern the life sciences and have avoided embedding its policy making in wider ethical debates. This does not mean China's science governance is not authoritarian. Rather its authoritarian element is exhibited not so much through setting up policies, but through enacting blanket bans and taking policy U-turns at times of controversy so as to silence criticism (Zhang, 2017). For example, as demonstrated below, despite first being celebrated domestically as 'a big step forward' for science, China effectively banned hybrid embryo research within days of it meeting international disapproval. A lack of regulatory clarity and consistency still looms over Chinese scientific practitioners and the biomedical industry, causing disruption (Zhang, 2017). As such, it is arguably more accurate to describe Chinese policy rationales as 'passive' rather than 'permissive', for policy is less about taking a clear stand on contentious science than about waiting to be pushed into adopting a position through crisis

(Miller, 1996). Consequently, as some have rightly observed, distinct from Soviet style central planning, China's modern science and technology system has always had 'a strong local element' in its planning and organisation (Gu, 2001: 206–7). This has led to disparities in the consistency and quality of governance across difference regions. Such divergence is further aggravated by an elite-focused funding strategy.

For much of the past 50 years, government funding has been a significant and sometimes the only source for cutting-edge research in the life sciences (Zhang, 2014). Given limited research, China has taken an elite-centric approach in allocating its investment. This can be seen from three parallel strands of government funding. One is the concentration on key national projects. The founding of the High-Tech Research and Development Program (better known as the 863 Program) in 1986, and the launching of National Basic Research Program (the 973 Program) in 1997 established two main national funding platforms that select and support ground-breaking projects that could give China a leap if not a lead in global science. The rationale for establishing global 'winners' was reinforced by the government's second strand of funding decisions, which focused on universities. In 1995, China launched the 211 Education Project with the aim of building more than 100 universities (now 116) for the twenty-first century as hubs for the next generation of human resource (State Council, China, 1995). This was followed by the 985 Education Project which supported 39 of these selected institutions to become world-class research-led universities (State Council, China, 1999). This has led to visible gaps of teaching and research capacity among Chinese higher education institutions. In her study of publication strategies among Chinese universities, Subrina Shen (2016) found that resource disparity has created a secondary gap in research cultures within the scientific community. Researchers at elite universities could afford to respect an individual lab's autonomy and had a more competitive and negotiated collaboration between labs. Authorship of papers was normally accredited according to one's actual scientific contribution. In contrast, provincial universities, which on average have less than one-sixth of the financial resources as elite universities, resorted to a hierarchical control of limited resources. An 'equal' and rotational authorship attribution is used to acknowledge long-term co-dependence between scientists, technical administrators and sponsors (Shen, 2016). A third strand of government funding lies in its efforts to reverse the brain drain. In 1992, the Ministry of Personnel and State Education Commission introduced efforts to create postdoctoral positions and attract competitive overseas PhDs. This was joined by the State Council's decision to give blanket 'immunity' to any overseas returnees for their criticism over the 1989 Tiananmen Protests (Zweig, Fung and Han, 2008). In 1994, the Chinese Academy of Science (CAS) established the One Hundred Talents programme to attract high achieving young scientists. This was followed in 1996 by the 'Spring Light Project' and in 1998 by the 'Changing Scholar Programme', both initiated by the Ministry of Education to offer short-term, flexible work arrangements. In the first decade of the twenty-first century, coordinated efforts between five Chinese ministries in the areas of education, science and technology, personnel, finance, and public security introduced a series of regulations to facilitate overseas talents' long- and short-term research exchanges with China (Zweig, Fung and Han, 2008). Marching into the second decade, it increasingly became common practice for various Chinese institutes and universities to conduct headhunts of their own, often targeting early-to-mid-career researchers with competitive publishing and funding records. In addition to public advertisements, institutions would directly contact individuals with 'settlement packages', which typically included a start-up grant, research lab facilities, competitive salary, and free or heavily subsidised housing.

There is nothing essentially wrong with concentrating national funding on selected areas or selected institutions or individuals. However, there are at least two factors that makes China's funding strategy highly elite-centric and, as we later show, encourages a perverse entrepreneurship in Chinese science. To begin with, its funding structure further polarises rather than democratises the scientific community. In the 1990s, the founding of the NSFC was considered a breath of fresh air, ushering in transparency and fairness due to its active engagement with international experts in its peer-review of grant applications (Zhu and Gong, 2008). But for the majority of funding decisions, government officials or academicians have significant influence over decisions on the content and the personnel of public-funded projects (Cao and Suttmeier, 2001). Until recently, it was not uncommon for key national projects to be 'commissioned' to politically and scientifically well-connected individuals rather than through a fair or transparent process (Cao, 2004; Zhang, 2012b). Scientists' social and political capital, such as knowing the right person or having international prestige, remain as important as having a well-formulated grant proposal. Overseas returns often struggle to secure national and provincial grants in the first few years of their settling back in China for their lack of personal connections with scientific and political elites.

A more worrying and less discussed aspect of government funding strategies was the impact of a utilitarian rationale that viewed investments outside of elite projects as a 'burden'. During the 1980s, CAS relied on 'One Academy, Two Systems' (Cao, 2004; Cao et al., 2018). Instead of government funding, CAS's survival relied on having a creative way to establish non research related affiliated businesses such as 'retails shops, restaurants and building cleaning services' (Ma, D., 2019: 382). A popular joke at the time was that 'those who work on making bombs live poorer than those who work on making tea-eggs'. During this time, in the public's view, it was difficult to differentiate CAS as a scientific institution from CAS as a business (Ma, 2019). In 1998 when the government designated CAS as the pilot institution for its Knowledge Innovation Reform, science was once again at the centre of CAS's agenda (CAS, 2009). In 2011, the government pushed forward its reform on public institutions with a focus on economic efficiency (State Council, 2011). Worried about its own position, CAS launched a proactive internal reform where institutions were regrouped into commercial-oriented, global-research-competitiveness-oriented and development-and-sustainability-oriented categories (Cao et al., 2018). As Xi Jinping's leadership further underlined the value of demand-led and problem-led research and innovation, 'rapid breakthrough and effective technical solutions' became the guiding principles for China's R&D investment (Ren and Wan, 2020). In such a rapidly changing context, it seems that even CAS struggled to maintain government recognition of its relevance. As part of the government's initiative to economise spending in public institutions, China's Ministry of Human Resources and Social Security (2017) promulgated a national policy that encouraged technical personnel in public institutions to suspend their employment and join industry or start-ups. Higher education and health care were the two main sectors targeted in this reform. Following the call of this 2017 Directive, Jiankui He was one of many young professors who were encouraged by their institutions to take long sabbaticals and explore marketisation opportunities of their research. While such reforms were said to promote horizontal mobility of talent between sectors and encourage healthy competitiveness, they were also criticised, especially in health care, for having effectively become an excuse for local authorities to 'dump the burden' (*shuai baofu*) of public health provision to local communities (*Beijing Youth Daily*, 2016).

In short, the Chinese government's interest in science lies in it being instrumental in speeding up socio-economic development and enhancing China's political leverage abroad. As such, its funding strategy is elitist. The Haidian District in northwestern Beijing, for example, arguably concentrates more funding and research opportunities than some provinces. As such, overseas returnees have been advised 'to first establish roots in Beijing before migrating to other cities' (Jia, 2020). Yet far from a common misconception that science is centrally organised and regulated, there are huge regional disparities in research capacity and research norms. For the central government, good science should be able to contribute to economic and political glory rather than cause tensions. Anything that distracts from economic growth, such as social debates on new technologies or supporting local universities, could be considered as an administrative or financial burden and should be sorted through the market or by local authorities. What is perhaps most revealing is that even CAS as a national academy needed to take the government's expectations on economic returns seriously for its own financial survival. It is no wonder that some have insightfully noted that Singapore's biomedical community was a result of 'a concerted state initiative' whereas China's life sciences were a success story of research groups' bootstrapping 'despite the state' (Fischer, 2018: 286). As the government was reluctant to risk investing in research with no prior guaranteed impact, scientists needed to demonstrate their relevance to either economic or political agendas in order to gain institutional support. Studies have found a pervasive anxiety among Chinese scientists to 'survive and thrive' in the domestic research ecology, regardless of whether they were based in an elite or secondary level institution

(Shen, 2016: 673). This point is best demonstrated by the case of the BGI Group.

From Beijing to Shenzhen: BGI and scientist–state relations

Huanming Yang and BGI are two names that everyone who researches the life sciences in China will eventually come across. BGI, formerly known as the Beijing Genomic Institute, affiliated to the Chinese Academy of Sciences, was the only institute from the developing world that contributed to the HGP. It was no small contribution either, as the institute alone mapped 1% of the human genome. The story of how that 1% became China's task was recounted to us by many over the years, through interviews, at dinner tables or during a casual chat in a taxi. The stories often vary slightly, each with a different reading of the event. Some did not remember the date was 1 September 1999; some mistakenly told us it was the Fifth International Strategy Meeting on Human Genome Sequencing in the US, while it was actually in Cambridge in the UK. There was also different emphasis on the praise and criticism Yang received immediately after his return to China from the conference. Yet, despite the conflicting narratives, everyone remembered Yang as a 'nobody' before the HGP meeting. It was also common knowledge that at the time, Yang did not have the Chinese government's, or even his home institution's, prior consent or knowledge on his pledge. More importantly, one detail remains strikingly consistent: on the day of the Cambridge meeting, Huanming Yang 'jumped onto the stage' and made the sequencing pledge on behalf of China – for through a patriotic lens, Yang could not fathom how China could be left out of the global initiative. We found the repeated use of the verb 'jump' (*tiao*) interesting, as in Chinese the verb could be interpreted as either a reflection of Yang's determination, as a dramatic turn of events, or as an act of impudence with the risk of making a fool of oneself. In fact, Yang's appearance at the meeting was not an act of impulse but consisted of a 25-minute presentation to the conference delegates on the feasibility of his institution to carry out the proposed 1% sequencing (Liu and Feng, 2004). In short, years after putting gene sequencing on China's national research agenda, Yang and his colleagues considered it easier to persuade the international community rather than their own government of his institute's ability to help sequence the genome. Yang's daring act in 1999 has become a critical event, not only to his own career, but also to China's genome research. In 2007, BGI's headquarters moved from China's political capital in Beijing to its southern entrepreneurial hub Shenzhen. While Yang remains an academician of CAS, the move to Shenzhen signalled a new level of BGI's independence: that it had become an empire of its own and could afford distancing itself from political and academic hierarchies in the north. In addition to its domestic branches, currently in 8 cities, BGI has also opened offices in the US, Europe and Japan (www.genomics.cn), providing essential research services to an expanding list of agencies worldwide.

Arguably the tale of Yang's critical 'jump' has acquired a life of its own. BGI's impact lies as much in what it did scientifically as in the conflicting views it provoked within China. Some considered Yang and his team as underserving of their fame. For them, Yang and his partners are corrupted technocrats who turned government funding into a private venture, continuing to benefit from 'low-skill' lab productions of gene sequencing. Others considered Yang as a dedicated if not misunderstood scientist who had rare foresight and talent in opening up an entire field for China, and whose only fault was his forced engagement with the business side of things due to a lack of government support. Unlike other controversial figures such as Xigu Chen or Jiankui He, who quickly receded from public spotlight, Huanming Yang has been a vocal figure in the life sciences for more than three decades. Apart from being a founder of Asia's biggest genomic empire with a growing number of employees, he is also active in bioethical debates and has supervised many doctoral and master students. So while we want to be cautious not to overstate Huanming Yang's celebrity status, it is also our experience that when talking to scientists and ethicists in China, it seems that every one knows Huanming Yang or knows someone who knows about him. Different tales about BGI's success continue to be told for their symbolic meaning, reflecting a deeper ambivalence and anxiety about having a scientific career in China. This makes BGI particularly instructive about scientist–state relations.

In this section, we recount both sides of the narrative. Underlying their apparent contradiction, they both point to the fragility of scientist–state relations in China, where scientists' status is contingent upon their ability to 'prove' their worth. Viewed in this way, Huan-ming Yang and BGI's eventual move from Beijing to Shenzhen are emblematic of an entrepreneurial mentality that both fuels and plagues the development of Chinese life sciences.

One of the authors, Joy, originally thought she would meet Huanming Yang in November 2006 in London at the kickoff meeting of BIONET, an EU-funded research project on the ethical governance of biological and biomedical research led by Professor Nikolas Rose, then at the London School of Economics and Political Science. Yang was one of the 21 partners. But due to a scheduling conflict, he decided to send one of his most promising master students as a delegate instead. For some, this unconventional choice may appear to be contemptuous, but it was actually a practical decision as the student was also head of a research office in BGI and participated in ethics decision making at the institute. This gives an insight into BGI's institutional culture and perhaps also Yang's personality of not being blinded by convention. In fact, BGI was known for its vibrant young work force. In 2010, *Nature*'s Editorial (2010: 7) marvelled at how BGI's in-house training had turned many under-graduates into 'tremendously experienced' researchers. Many of these budding scientists, the editorial noted, chose to 'forego conventional postgraduate training' for the chance to be part of BGI's self-directed research initiative that promotes both technical skills and scientific creativity. The piece argued that 'the BGI model may be worth serious consideration' as an alternative to lengthy Western-style postgraduate training. It was not until a year later, when Joy visited BGI's Beijing headquarters that she first met Yang in person. After seeing how Yang critiqued two of his PhD students' research plans and strategised market expansion with his colleagues over lunch, she got a sense of how he could so quickly establish a larger than life strong presence.

'But he is an opportunist!' one scientist in his late forties from a different department in CAS tried to explain. For this interviewee, Yang's success was built on 'hijacking' the government's financial backing. Upon Yang's return from the 1999 human genome sequence conference, it was said he was 'scolded' for his reckless behaviour representing China without institutional approval. But as global media started to celebrate China as the only developing country to contribute to the project, the Chinese government eventually went along with his pledge and invested 22 million RMB (approx 2.7 million US dollars at the time) to support BGI (Liu and Feng, 2004). 'Do you know the opportunity cost of that investment?' the interviewee added. 'It could support a CAS institution for a whole year! Imagine how many projects that could support.' One only needs to be reminded of CAS's 'One Academy, Two Systems' struggle in the 1980s and early 1990s, discussed in the previous section, to understand the interviewee's sentiment. This professor was not alone in thinking BGI was unfairly rewarded by the government. Eminent neurobiologist Rao Yi (2012) flatly called Yang a 'gangster entrepreneur' who took advantage of a 'massive amount of funding and cheap labour from China', as well the credibility of CAS's affiliations, to boost his private business. It does not help that Yang's main partner, the current owner of BGI, Wang Jian, often flaunted the label 'bandit scientists' (*kexue tufei*) (Hexun, 2017). But he wasn't referring to robbing his peers of their research opportunities or cheating government funds. On the contrary, he referred to their experience of not being able to get government recognition, let alone funding, through normal channels despite having a 'world class scientific project' (Liu and Feng, 2004). To carry on with their scientific projects, they had to act like 'bandits' to look for support in the 'wild', that is, seeking any support they could get from outside of the academic system.

Thus, the popular narrative that BGI hijacked the government could be turned upside down. According to Yang and Jian Wang, BGI had always been an independent research institute and its success lay in proactively building government confidence in a new research area. In 1997, Yang and Wang, both overseas returnees, saw the huge potential in the field through the HGP and applied to CAS for funding. CAS approved of this project but apart from 500 square meters of empty office pace, 'funding, people and equipment were all left blank' (Liu and Feng, 2004). After exhausting funding avenues from CAS, MOST and NSFC, as well as local governments and industries, they were forced to set up BGI so as to seek alternative funding (Fischer, 2018; Hexun, 2017). Wang described the early days: 'The state didn't recognise (the importance of our research), industry didn't either, no one responded to me' (Ye, 2013). For Yang, the government's early reservations were due to 'decision makers' lack of confidence', for even shortly before the Cambridge meeting the Chinese government was planning to pay 1.84 million RMB to subcontract its domestic needs of gene sequencing to foreign companies (Liu and Feng, 2004). Yang and Wang used personal connections and secured seed money to continue their sequencing task. Critical early financial support came from the mayor of Yang's hometown, Leqing, a city in Zhejiang. The mayor was moved by Yang's personal charm and agreed to loan him 11 million RMB, 'despite not understanding a thing about the project' (Fischer, 2018; Liu and Feng, 2004). Internally, BGI promotes this experience, rather than the later success, for Wang still emphasises to his employees that besides a professional qualification, good scientists need to have the courage and the skill to 'beg' (*yaofan*) when their project demands it (Ye, 2013). To some extent, Yang and Wang's perceived lack of appreciation for government support was understandable if one takes into account that Yang spent 70,000 US dollars (approximately half a million RMB) of his own savings to sustain his lab. Despite years of rejection from national funders, within three months of the Cambridge conference, MOST approved a 40 million RMB research budget for HGP-related research, of which 22 million was for BGI (Liu and Feng, 2004). But BGI's other projects remain reliant on various local government funding and they also have risks. When local officials change posts, a previously agreed upon grant may be cut short (Ye, 2013). When on 5 April 2002, BGI's publication of the draft sequence of Asian rice variety *Indica* made *Science*'s cover story, it left BGI 'deep in debt' (Ye, 2013; Kennedy, 2002). The real change in fortune for BGI, as Jian Wang recalled, was 'when Bill Clinton said thanks to the Chinese scientists who also contributed to the Human Genome. Jiang Zemin (then President of China) began asking "who did it?!" Then he came down to visit us' (Fischer, 2018: 227).

Both narratives of BGI's success illustrate how scientists who felt they were left out needed to compete for government recognition. More importantly, such recognition was not necessarily gained through informed decisions on science but on the ability to boost China's political or economic image. In the views of many, before 1999, Yang was a nobody, just another colleague at CAS. Looking upwards from the grassroots, an authoritarian government appeared to act more like a venture capitalist, viewing science as a tool to 'increase the forces of production' rather than to 'explore the unknown' (Miller, 1996). But unlike many of his CAS colleagues, Yang 'jumped' into scientific stardom for he rightly gambled on his project's ability to shape China's global image and thus change the government's mind. For some, Huanming Yang's cunning provides an enlightening role model to navigate through China's science system. For others, their personal grudge against BGI is rooted in a deeper frustration of the scientist-state dynamic in China: entrepreneurship rather than scientific ingenuity is the shortcut to success. It validates a 'bandit' spirit where reckless behaviour is wrongly rewarded as intrepidity.

BGI's long struggle for government recognition also helps to explain its deep ambivalence towards the role of government and government-backing in science. On the one hand, the aura of having the central government's endorsement also lightened BGI's path to commercial success. In 2007 BGI relocated to Shenzhen as a self-sufficient institution. A decade later, BGI made its initial public offering at the Shenzhen Stock Exchange. Shortly after its debut, shares surged to the 144% limit of the initial offering. There can be little doubt that an annual revenue of 250 million USD and a projected annual growth rate of 20–25% were crucial factors in its success (Xinhua, 2017). Yet, on the other hand, BGI was also not shy to use its hard-fought research freedom to publicly snub the establishment's ineffective ways of delivering science. According to 2019 *Nature Index*, BGI was the most productive genetic research institution in Asia in terms of academic articles, ranking 15th in the world, higher than elite Western institutions such as Johns Hopkins and Stanford University. As of December 2020, BGI had published 349 articles in *Nature*, *Science* and their affiliated journals (BGI, 2020). BGI specifically highlighted on their official website that 93% of academic papers BGI published had come since its relocation to Shenzhen. BGI's flaunting of its 'post-Beijing' (or 'post-CAS') productivity is a celebration of its hard work, but it is perhaps also symbolic of a new era of how high-level scientific research can and is being done.

It is perhaps of little surprise that 19 years after Yang's daring act on the world stage, another up-and-coming Chinese scientist, his Shenzhen neighbour, Jiankui He, planned on making another stunning announcement at an international summit on gene research, with the expectation of a heroic return to China. Although He cited his Stanford postdoctoral adviser Stephen Quake rather than

Huanming Yang as his role model of 'being a scientist and an entrepreneur' (Li and Yan, 2018), their approaches bear a striking resemblance. There is also nothing intrinsically incompatible between science and entrepreneurship. The 'demands for scientists to generate *impactful* research are ubiquitous' (McLellan, 2020: 1, original emphasis). Science historian Philip Mirowski (2002) would go as far as arguing that the 'history of technology' can be integrated with the history of economic ideas. In the US, the introduction of both the Bayh-Dole Act and the Federal Technology Transfer Act in the 1980s formed a global foundation of translating publicly funded research into patentable applications (Mirowski and Sent, 2002; Oliver, 2004). But crucially the integrity and legitimacy of science depends on 'how states recognise who is an expert' (Jasanoff, 2012: 11).

It is on seeking state recognition that we see a connection between Huanming Yang and Jiankui He. The line between audacity and recklessness in entrepreneurial scientists can be a fine oe. In the case of Huanming Yang and Jiankui He, we particularly want to underline the repeated pattern of 'science through PR' and the shared perception that acquiring global recognition as a *shortcut* to get national attention (Mallapaty, 2021a). This peculiar rationale provides an essential insight into China's scientist–state relations and its underlying risks.

At one level, it is not difficult to understand how this entrepreneurial mentality in science is nurtured precisely by the Chinese government's positioning of itself as the client rather than sponsor of science (see previous section). Its limited tolerance for experiment failure or low-returns often skew scientists' vision of what counts as success and how to achieve it (Han and Li, 2018; Suo, 2016). One only needs to be reminded that while BGI's success in sequencing the *Indica* rice genome left it in debt, it was Bill Clinton's simple praise that drew the Chinese president's attention. As such, Jiankui He's dual-language promotion videos and his bold plan to reveal his research globally while shortcutting domestic bureaucracies made sense. He expected that global attention, rather than science itself, would boost him into the league of scientific elites. It was not too far-fetched to believe that political glory and international recognition could help bend the rules at home. After all, as the section on state-ethicist relation demonstrates, respecting the rule of law and bioethical norms are still ongoing initiatives in China. From the point of view of a government pressured to forge socio-economic progress on multiple fronts, 'shopping for success' in domestic science may make sense. As discussed further in Chapter 5, China kicked off its 'mass entrepreneurship and millions to innovate programme' in mid-2015 and its latest encouragement of academics joining with industry or creating their own start-ups stated in the Fourteenth Five Year Plan, are in line with this 'shop for success' scientist–state relation (Mallapaty, 2021b).

However, at another level, China is also at a pivotal moment in recognising that such scientist–state relations are unsustainable. For as contemporary life science sprung out of traditional institutions, the role and leverages the state occupies are also becoming ever blurred. Around 2000, when the development of China's life science began to accelerate, despite China's lagging behind in ethical regulations, public funding offered a lifeline for research in China. The government still has effective control over and sufficient knowledge of domestic research activities. Today, BGI flaunts its post-Beijing productivity. Jiankui He accomplished a world-first without bothering to inform the local authority. With intensified industrialisation and privatisation of cutting-edge science, it seems that the state could be a new elephant in the room. How could the Chinese government maintain its relevance to, and authority over, these self-sufficient ventures, especially as they gain significance to China's economic success as well as its political image? One could say we are worrying unnecessarily. According to the Chinese financial think tank, List Company Research Institute's 2019 analysis, the marketisation of Chinese biomedicine remained at least ten years behind developed countries. Yet in investment hubs such as Shenzhen and Shanghai, there are already plenty of venture capitalists looking for bioscientists. For a country that once relied on funding as an administrative tool to control what science was done and by whom (Cheng et al., 2006), where should its authority now lie? This is a question that Chinese bioethicists have been trying to address.

From 'Wild East' to 'Wild West': state–ethicist relations

Bioethics is a relatively young profession in China. The country's rise in global science was also the process for bioethicists to gain recognition and socio-political space. The modern concept of 'bioethics' was first introduced into the Chinese language in 1987 through Qiu Renzong's monograph of the same name (Xu, 2010). But at least for the first half of its development, Chinese bioethics remained an academic discussion, especially in relation to ethical issues involving clinical practice. Upstream engagement with policy and science innovation were rare. Only in 2000, when the Chinese Ministry of Health (now known as the National Health Commission) established a ministerial-level Medical Ethics Expert Committee did bioethicists have a (semi-)official role in policy making. Although, even then, most were consultations as a matter of formality rather than sincere requests for advice (Zhang, 2012a).

The tricky balance for Chinese bioethicists is to establish their influence in science policy by demonstrating their added value to national competitiveness while not being seen as a threat to the technocrats within the Party. To be seen as a constructive partner to science is a concern for bioethicists around the world. But this is perhaps more important in an authoritarian regime where there was little tradition of sociological debate and a heightened suspicion of any critic who appeared to slow things down. More importantly, there are at least two things that the bioethical community sees as essential to their professionalisation, yet that are at odds with the government. First, contrary to the government's post-hoc pragmatism, which often yields vaguely worded position papers in lieu of substantial policies, since the late 1990s, Chinese bioethicists have been calling for clear wording and better enforceability of regulations (Li, Ji and Wang, 2011; Qiu, 2012). To be sure, this is key to normalising ethical compliance in research practice. A second aspect, as this section demonstrates, is that Chinese bioethicists are caught between their professional preferences for 'global dialogue' and the stance expected from the government to defend the 'Chinese characteristics' of science.

In what follows, we demonstrate how bioethicists have continuously sought a more valued position in national policy making through key events. We divide the past twenty years into three phases, each roughly consisting of seven to eight years. Naturally the phases are not mutually exclusive. But this structure helps us to see shifting strategies and priorities used by bioethicists.

Moreover, it allows us to see how the bioethical community evolves in response to critical events. We start with the 2001 Xigu Chen hybrid embryo scandal, which helped earn China its 'Wild East' reputation. It was also Chinese bioethicists' first major appearance on both the domestic and global science policy stages. Due to an under-development of domestic debate in this area, Chinese bioethicists resorted to borrowing Western (mostly British) standards and rhetoric, which promoted the institutionalisation of ethical guidelines in China. However, during the second phase (roughly between 2008 and 2015), a series of violations of ethical review procedures by American research teams evoked a domestic uproar against the 'Wild West'. This combined with repercussions from the hybrid embryo case to usher in a 'defensive turn' in Chinese bioethics, which became more sensitive and sceptical to an imbalance in power in global policy dialogues. The beginning of the third phase starts with stem cell scientist Junjiu Huang's CRISPR experiment on non-viable human embryos in 2015. The case bears many similarities to the 2001 hybrid embryo case, yet Chinese bioethicists' response was much more measured. This was also a turning point where leading bioethicists, through media interviews, began to downplay the term 'Chinese particularities' in favour of global connectedness.

In 2001 when the *People's Daily* reported Chen Xigu's creation of the world's first human-rabbit hybrid embryo, it was celebrated as 'big step forward' in stem cell research. The report also stressed that Chen or any other Chinese scientists will 'never make any attempt in researching reproductive cloning' (Zhang and Chen, 2001). Yet, in Western media, the phrases such as 'a morally bankrupt "Wild East" of biology' came to characterise the Chinese life sciences (Dennis, 2002: 334). Chen, a professor at Sun Yat-sen University, inserted a skin cell nucleus from a 7-year-old boy into a rabbit's denucleated egg (Zhang and Chen, 2001). At the time, embryonic research had spurred contentious international moral and religious debate. Yet bioethical discussion in the public domain was effectively non-existent in China and the majority of Chinese universities, Chen's included, had yet to embrace institutional ethical review processes. Thus, despite the fact that Chen's research did not violate any domestic regulation or international consensus, it immediately became 'the most controversial case at that time' (Abbott and Cyranoski, 2001; UNESCO, 2008; Zhang and Chen, 2001).

This 'Wild East' controversy stung Chinese bioethicists in two ways. At one level, it exposed an embarrassing gap of ethical governance in China. But bioethicists were quick to turn this into a positive force for inserting their voice into Chinese politics. Ethicists made a series of public statements re-emphasising China's firm commitment to developing stem cell research ethically (Wang, 2003). One month later, as a result of working closely with stem cell scientists, the Chinese National Human Genome Centre at Shanghai (CHGC) issued the *Ethical Guidelines on Human Embryonic Stem Cell Research (Recommended Draft)* (Doering, 2004; Ethics Committee of CHGC, 2001). The focus was threefold: 'to dissipate foreign misunderstanding', 'to support Chinese stem cell research with firm and sensible reasoning' and to establish a 'positive and orderly' research norm (Zhang, 2012a: 48). However, despite the fact that bioethicists were nominally consulted by the government, the policy drafts and comments submitted to the Ministry of Health went unheeded. When the national Ethical Guidelines for Research on Human Embryonic Stem Cells was promulgated in 2003, it mainly consisted of statements of internationally agreed principles rather than operable instructions on how ethical oversight could occur (MOH and MOST, 2003). It was also known as 'an ethical guideline that excludes the ethicists' (Xiao, 2004), for among the prescribed 'essential' members of an ethics committee were biologists, medical experts, lawyers and sociologists; in other words, ethicists were nowhere mentioned in the national 'Ethical' Guidelines'. In the same year, Huizhen Sheng, Chief Scientist for the national 973 Program, tried to publish her findings on human–rabbit hybrid embryos with proper institutional ethical approval. Yet her paper was still rejected by prestigious journals, such as *Science* and *Proceedings of the National Academy of Sciences* (Fox, 2007; Mandavilli, 2006). In the end, she was only able to publish in *Cell Research*, a China-based English journal (Chen et al., 2003). The reception of Sheng's research was mixed. To calm international criticism, Chinese authorities moved to effectively ban such research and shut down the debate.

Sheng's experience constituted part of the second sting. In the years that followed, international commentators started to develop differentiated views on the type of hybrid embryos Chen and Sheng created. The reason for the shift was that such research allowed the creation of stem cell lines that were effective for clinical research while reducing the need for human eggs (Zhang, 2012a). After a comprehensive and open review that drew on expert and public opinion, in early 2007, the UK's Human Fertilisation and Embryology Authority (HFEA) endorsed scientists conducting cytoplasmic hybrid research, which was identical to Sheng Huizhen and Chen Xigu's previous studies (*Lancet*, 2007). When the UK's first hybrid embryo was created on 1 April 2008, the Chinese scientific community realised that the government's hastened ban had cost China's lead in the area. Funding had long been discontinued and research groups dissembled (Ji, 2007; Walsh, 2008).

The double sting of the Wild East episode demonstrated to China the power of bioethics. It was also a powerful demonstration of how simply following Western discourse was not sufficient, and potentially even costly, to China's fragile but burgeoning life sciences. It was a sobering moment, too, for bioethicists, as even with mounting international criticism of China's research ethics gap, they still struggled to get authorities to take their advice seriously. In the aftermath of the hybrid embryo controversy, the Deputy Director of CAS, Zhu Chen, called for increased investment to build capacity in bioethics (CAS, 2004). Within the bioethics community, two modes of doing bioethics were distinguished: the kite-flying mode, which followed a principalist approach, seeking a universal framework that could guide all scientific practices; and, the 'bicycle riding mode', which proposed field experience in order to identify contextualised ethical solutions and justification for research (Qiu, 2012).

While ethical oversights were yet to be taken seriously within China, during the second phase, Chinese bioethicists assumed two roles. On the one hand, they assumed the role of informal ethical patrols who were often sceptical, at times even 'paranoid' of international scientific collaborations. One typical example was recorded by Haidan Chen in her article on China's Kadoorie Biobank. The article began with a vivid illustration of the UK Medical Research Council's CURE (China-UK Research Ethics) Workshop in 2009. When an Oxford-based Chinese researcher started to present the Kadoorie Biobank study of 0.5 million adults in China, Chinese bioethicists in the audience immediately fired off a series of accusatory questions:

> 'Did you ask participants to sign informed consent?', 'Did participants know what their blood samples would be used for?', 'Who approved your research?', and so on. Chen was peppered with so many questions about his research ethics that he was scarcely able to describe what the project had accomplished. A Chinese bioethicist sitting next to me spoke angrily to no one in particular: 'They must have bought these blood samples from blood stations in China at a low price, otherwise, how could they obtain such a huge number of samples? (Chen, H., 2013: 322)

Haidan Chen depicted the caricature of Chinese bioethicists to underline a heightened awareness of informed consent and benefit sharing from the Global South. The Kadoorie Biobank was established as a high-level collaborative effort between the Chinese Centre for Disease Control and Prevention (Chinese CDC), the Chinese Academy of Medical Sciences (CAMS) and the University of Oxford. Although it may not have been her intention, the contrast between the 'professional' nature of Kadoorie and Chinese bioethicists' imagined fight against 'rogue projects' made Chinese bioethicists out to be some kind of neurotic Don Quixotes.

Yet this 'paranoia' may not be so hard to understand as it reflects a deep frustration among Chinese bioethicists with ethical compliance within China. At the time, a decade into the twenty-first century, China was still limping behind on embedding research ethics into institutional practices. In the absence of penalties for non-compliance and training support for setting up ethical review boards, many institutions simply rubber stamped research after it had been conducted but before publication (Huang and Li, 2012). Having a high-profile collaborator, such as the Chinese CDC, did not necessarily mean ethics were properly adhered to.

In fact, three years after the conference cited above, the Chinese CDC was implicated as a partner in another damaging 'Wild West' case. In 2012, Tufts University professor Guangwen Tang was criticised for conducting a Golden Rice experiment with 72 primary school students in Hunan province. The project had the China CDC, Hunan CDC, and Zhejiang Academy of Medical Sciences as collaborators and involved serving the experimental rice to the students

(Qiu, J., 2012). In addition to not obtaining ethical approval in Hunan, the bigger problem was that the parents of the school children were not fully informed. Key words such as 'genetically modified crop' and 'golden rice' were never mentioned to the parents. Within months, the government sacked the three officials involved in fabricating research documents (Qiu, J., 2012). After a year's investigation, Tufts University also apologised and its principal investigator, Tang, was barred from human subject research for two years (Enserink, 2013).

For Chinese bioethicists, this case exposed the 'urgency to enhance ethics review capacity' and the 'urgency to enhance supervision over ethical review committees' (Qiu and Zhai, 2013). China had already introduced the trial version of the Measures for the Ethical Review of Biomedical Research Involving Humans in 2007 with the intention of embedding ethical scrutiny into the design and delivery of research. However, in the Golden Rice case, Chinese officials who fabricated the ethical approval and hid key information from the consent process said they merely 'wanted to save time and push the project through, and … did not realise how serious the matter was' (Qiu, J., 2012).

While bioethicists have had moderate success in pushing Chinese regulators to promulgate ethics review regulations, they were still not being taken 'seriously'. The call for deepening institutionalisation of ethical oversight was echoed in a number of articles in high impact Chinese journals (see Li, N., 2013; Teng and Feng, 2013; Tian, Yuan and Ouyang, 2013). As a result, in 2016, the ministerial-level Medical Ethics Expert Committee was further upgraded to the National Medical Ethics Committee serving the newly formed National Health Commission (National Health and Family Planning Commission, 2016). A new task of the National Committee was to provide training, supervision and guidance on capacity building at provincial ethical committee level so as to facilitate further down-stream ethics training.

The most recent critical event for Chinese life sciences as well as for Chinese bioethics is the CRISPR controversies between 2015 and 2018. While Jiankui He is now the better-known scientist, it was actually another Chinese scientist that made the breakthrough in applying CRISPR to editing human embryos. In 2015, Junjiu Huang at Sun Yat-sen University published his results on using CRISPR/Cas-9 technology to edit the hemoglobin-B gene in 86 discarded IVF embryos donated for research by couples at an IVF clinic (Liang et al., 2015).

Both the publication and its international reception of Huang's paper brought peculiar familiarity with Huizhen Sheng's hybrid embryos case. Similar to Sheng, Huang obtained ethical approval from his university ethics review board. But his work had already been rejected by leading journals, such as *Science* and *Nature* due to ethical concerns. Eventually, the team took a similar decision as Sheng by going with *Protein & Cell*, an English journal based in China (Cyranoski and Reardon, 2015). The paper 'split the scientific community' and 'spark[ed] epic debate' (Cyranoski and Reardon, 2015; Kaiser and Normal, 2015). Some leading figures in the field, such as George Church and Tom Daley cautiously praised Huang's approach as 'noteworthy' and stated that there long had been a consensus that research on non-viable embryos was acceptable. Others condemned such research and, in response, watchdogs of human genetic and reproductive technologies called for a halt to such experiments (Center for Genetics and Society, 2015). Given that the paper was published two days after submission, some Western scientists dismissed it as not peer-reviewed and thus not valid (Kaiser and Normal, 2015). Yet the editors of *Protein & Cell* explained that they had fast-tracked the paper with the benefit of reviewers' comments from *Science* and *Nature*. As such, and in a world where labs around the world are competing for results, two days, in the words of the journal's editor-in-chief was 'very long' (Chen, 2015).

Chinese bioethicists' early involvement in formulating a national response was a significant factor in shaping a more measured reaction in 2015 as compared to the hybrid embryo case more than a decade earlier. In particular, Xiaomei Zhai, Office Director of the National Medical Ethics Committee at the time, was called on by the Ministry of Science and Technology and Ministry of Health to draft a response. Zhai later told Joy that the views among decision makers were also split: some had insisted on adopting a similar tone as in the hybrid embryo case and denouncing Huang's research for not having had wider debate about its ethical implications. The rationale was to signal a warning to other ethically contentious experiments around the country. Others, such as Zhai, saw it as a matter of following a rule of law. As Huang did not violate any national or international regulation or scientific consensus, as a bioethicist, she felt China needed to back and explain Huang's research in the face of global scepticism. In fact, as we've seen so far, establishing an orderly research culture through good enforcement and compliance is considered crucial for ethical governance. 'It was a close call!' Zhai later told Joy in Beijing. The ministries eventually adopted Zhai's advice. Both the official stance and media communication emphasised that the publication would 'sound an alarm' (Cyranoski and Reardon, 2015) and that the findings from Huang's research, as stated in the original paper, 'underscore[d] the challenges facing clinical applications of CRISPR/Cas9' (Liang et al., 2015). In January 2016, eight months after Huang's paper, the UK's HFEA granted approval for gene editing of human embryos to a group of researchers, led by biologist Kathy Niakan, at the Francis Crick Institute (Grush, 2016).

Zhai was not alone in seeing the CRISPR controversies as another case to convince the Chinese government to give bioethicists more meaningful roles in their strategic planning. As the profession matures and expands, it no longer wants to consist of firefighters who are brought in only at times of international crisis. Following the Jiankui He CRIPSR baby scandal, Chinese bioethicists quickly warned of the re-emergence of the old 'Wild East vs Wild West' rhetoric (Zhai et al., 2019). Citing examples of the Golden Rice controversy and Jiankui He's close contact with elite American academics throughout his experiment, Chinese bioethicists pointed out that such antagonist juxtaposition was counter-productive, for Chinese and Western bioethicists were not on opposite sides of a war, but were confronting the same problems (Zhai et al., 2019). A group of Chinese ethicists published a list of policy recommendations in *Nature*, calling on China to reboot its post-hoc pragmatic governance regime, ranging from institutionalising bioethical education to consolidating ethics compliance (Lei et al., 2019). To some extent, bioethicists voicing their hopes and commitments in *Nature* is not that dissimilar to Huanming Yang making a pledge at the HGP conference. Having such policy recommendations published in a leading international journal is arguably more powerful than domestic policy submissions. As bioethicists struggle for political recognition, they expect global discussions will have the same boomerang effect on the Chinese government.

Bioethicist is a new entrant to Chinese politics, which the Chinese government and research institutions still need to get used to. For many in the West, it is easy to see Chinese bioethicists as mouthpieces for the Chinese government. Their critical view towards Western governance often appears to be overly sensitive and defensive at international venues, such as in the China–UK conference vignette we cited, but a chronological review of the discipline's evolution in China renders a more nuanced view.

Problems over China's global image have provided avenues for bioethicists to fight for an independent and respected role in policy making. There has been moderate success. While the 'Wild East' notoriety helped to give bioethicists a seat at the policy table, the 'Wild West' exploitation helped to push forward the institutionalisation and capacity building of bioethics. In contrast to the exceptionalism promoted by the Chinese government, bioethicists also expressed strong resistance towards the 'quite powerful' force within China to break away from international standards in favour of 'cultural differences' (Tatlow, 2015). In July 2019 China approved the proposal to set up a National Science and Technology Ethics Committee. This was compared to Nuffield Council in the UK and the President Bioethics Council in the US. Bioethicists in China have campaigned for such a move for many years. The committee was formally established on 21 October 2020 (Sun, 2020). However, the list of members has not been made public. While at least four bioethicists are said to be on the committee, some cynically described the committee's composition and function as being 'mysterious'. In other words, bioethicists still seem to have a long way to go to sustain meaningful leverage in Chinese politics.

From enthusiasts to sceptics: science–public relations

To date, apart from studies on grassroots movements and urban protests against genetically modified organisms (GMOs) and environmental pollution, there have been few empirical studies on how the Chinese public have been involved in and shaped China's science policy. This absence of literature corresponds to an absence of a culture of public engagement or even the need for it in China (Zhang, J. Y., 2018b). For a large part of modern Chinese history, the government has been able to count on the public's enthusiasm and support for new technologies (Ouyang, 2003). Such optimistic views remain today. According to China's 6th National Scientific Literacy Survey published in January 2021, 87.5% of those surveyed believe that science would lead to a better future; 72.1% agreed that despite its side-effects, the good outweighed the negative in science; while 84.9% believe that the government should support scientific research, even when it may not bring immediate benefit (Wu, 2021).

Yet marching into the twenty-first century, the Chinese public also seem to be more sceptical and resistant to government directives on science. The growing middle class, in particular, have sought to have their interests attended to and their voices heard. Xiamen residents' protest against the government-planned construction of a PX chemical plant near residential areas in 2007 and similar urban protests in 2011, 2013 and 2014 signal the public's shift in attitude towards development plans made without consultation (Sevastopulo and Hornby, 2014). The public's demand for data transparency and an accompanying grassroots-led mapping of national water and air quality are societal challenges to the government's monopoly over the production and interpretation of scientific data (Zhang and Barr, 2013).

When public concerns were ignored, they made visible dents in China's scientific ambitions. In 2016, a national survey conducted by *Science and Technology Daily* and the Chinese Academy of Science and Technology for Development concluded that public opinion had a 'direct impact' on China's ability to translate lab research into industrialisation. In some cases, public views have limited the speed of research development. For example, since 2008, China has listed GM crop research as a 'national major project' in its mid- to long-term food security plan (National Development and Reform Commission, 2008). As the world's second largest holder of patents on GM technologies, China has been a leader in researching phytase transgenic maize, which is a crop that could increase efficiency in animal feed and reduce farming-related pollution as well as carbon emissions (Ministry of Agriculture, China, 2010). In 2009 the maize was China's first transgenic crop to be issued a bio-safety certificate. China is the world's second largest producer of maize. At the time of approval it was expected to be a huge impetus for China's global competitiveness as well as sustainable farming (Biotechnology Research Institute of CAAS, 2010). However, due to mounting anti-GM sentiment among the Chinese public, the ambition of actually planting phytase transgenic maize remained locked away in research labs. While the Chinese public was generally pro-GM in the 1990s, since 2008, a series of food safety issues have stalled public uptake of GM crops. The 2016 survey also found that while public literacy of GM remained at the same level as in 2002, public support had drastically declined. Only 25.7% of those asked supported planting GM crops and only 18.9% were willing to consume GM food (Fu, 2016; Liu, 2016). In comparison, a decade previously, 65% of the public were willing to consume GM food (Huang et al., 2006).

The 2021 and the 2016 national surveys seem to depict conflicting images of the Chinese public: they seem to be both science enthusiasts and science sceptics. They advocate government investment in scientific research but seem opposed to the application of it. However, the conflicting attitudes become easier to comprehend if one takes into account that science–public relations in China have always been tinted by politics. The evolving attitude towards emerging science is not towards science per se, but towards the government's politicisation of science.

Chinese debates on modern science's social impact can be traced back to at least the 1920s in the aftermath of the May Fourth movement. For example, reformers such as Hu Shi and Ding Wenjiang argued in favour of empirical-based science against concerns raised by conservatives (Shen and Williams, 2005). Decades later, after the Cultural Revolution, students favoured natural science and engineering professions not only because of employment opportunities, but because individual merit was evaluated by seemingly universal criteria – that is, criteria independent of political change. In other words, for many Chinese students, scientific disciplines offered a sense of neutrality away from the madness of political factions and campaigns.

In the early 2000s, the government promulgated the *Law of Science Popularisation* and the *Action Plan of Improving Citizen's Science Literature* to cultivate a better-equipped human resource for its knowledge economy and to promote social uptake on new technologies (MOST, 2002; State Council, China, 2006). However, science education and the popularisation of science was framed to 'serve' industrial and agricultural production and 'communist ideology' (Wu and Qiu, 2012: 526). In the life sciences, compared to other Asian countries, Chinese civil groups (e.g. patient groups, human rights and religious groups) have little presence in policy making. Scientific agendas and their regulation are mostly led by core professionals, such as bureaucratic and scientific elites (Sleeboom-Faulkner, 2012). Even when scientific professionals engaged with the public, as comparative research on China's and the UK's public communication of biomedical practices has shown, experts in the Chinese media were often selected by the authorities for their 'policy compatibility' rather than acting as independent professionals who 'speak for themselves' (Ren et al., 2014: 372).

Since 2003, a flurry of nationwide scandals on adulterated or counterfeit food and Chinese authorities' various attempts at covering up or downplaying these events have shaken public confidence in the government's accountability. The tension was heightened with the Golden Rice scandal exposed in 2012. When it came to light that the Chinese CDC official involved in removing any mention of the rice being genetically modified for fear that it is 'too sensitive', victims' parents asked on Chinese national TV, 'If it's safe, why did they need to deceive us in this?' (Qiu, J., 2012). In the eyes of the public, when the government selects 'policy-compatible' scientists as their spoke persons, then science communication loses its ability to 'speak truth to power' and becomes instead 'power orchestrated truth'.

Government-related science scepticism can also be found in the hesitancy of some Chinese people to be vaccinated (Yang, Penders and Horstman, 2020). A series of manufacture and regulatory mishaps since 2006 led to low quality and even toxic vaccines being provided to hospitals, which caused severe side-effects and in some cases the death of children (Wen and Lipes, 2016). The gravity of public resistance towards vaccines was noted by the Ministry of Health in early 2010. However, despite government assurances, public scepticism prevented China from fulfilling its pledge to the World Health Organization (WHO) on eradicating measles before 2012 (Ma et al., 2010; MOH, 2010). An independent study has shown that Chinese anti-vaxxers are not against vaccines per se, but are worried about domestically produced vaccines being substandard or contaminated (Olesen, 2015).

While the public seems to be ever more critical about science, national surveys in 2007 and 2011 consistently show that 'scientist', as a general category, remains the most respected social group in

China, with the highest ascribed 'social credibility and trustworthiness' (He et al., 2012). However, public scepticism towards scientists speaking at public events also remains very high (Liu, 2016). Thus, in engaging with the public, Chinese scientists face a peculiar 'credibility paradox'. That is, an absence of visible institutional or official endorsements, conversely, would render them with *more* public credibility and lead to better conversations (Zhang, 2015). Thus, instead of contributing to formal channels of science communication, such as responding to queries in public events or writing popular science article for newspapers, many scientists prefer to act as 'informal risk communicators' (Rickard, 2011) for local civil societies or grassroots events (Liu, 2016). This echoes patterns observed in the UK, where public suspicion and rejection of science may not be a reaction to science itself, but to the perceived intentions and the 'behaviour, track-record, and trustworthiness of the institutions in charge' (Wynne, 1980; 2001: 454). When scientists' de-formalised their politically charged institutional affiliations and redirected their professional identity as fellow citizens, they seemed to (re)gain their credentials from the public.

The dual nature of science–public relations in China arguably says little about science or the public, but says a lot about the effect of the over-politicisation of science. China is hardly unique in having a national government actively shaping public opinion on contentious technologies (Bell and Hindmoor, 2009: 77–8, 86; Leong et al., 2011). Yet an authoritarian regime amplifies the impact of information control and consequent public doubts. Scientists' reorientation of their position as risk communicators, despite a financial and administrative monopoly by the government, is an involuntary restoration of a *public* reasoning of science (Wynne, 1980; 2001). This has helped to form a particular 'civic epistemology': that is, socio-politically grounded 'knowledge-ways' through which citizens 'assess the rationality and robustness of claims that seek to order their lives' (Jasanoff, 2004; 2005a: 249). The fact that public scepticism stalled the application of GM technology and caused China to miss a WHO vaccination target are good examples that even in authoritarian societies, 'civic' epistemology plays a tacit yet significant role in validating scientific knowledge. In 2019, Joy was invited to speak at the World Conference on Science Literacy (WCSL) organised by the Chinese Research Institute for Science

Popularisation, the equivalent of the UK's Science Media Centre. She was consulted on the wording of a memorandum that China was to sign with 19 countries (mostly in the Global South) in which the Chinese Research Institute was designated to take the lead in organising transnational collaborations to develop public engagement capacities (WCSLRT, 2019). It was a welcome move, symbolising China's acknowledgement of the need to take the public and public engagement seriously. However, for the scheme to work, perhaps the Chinese government needed to roll back its censorship and leave political space for public opinion.

From maverick to authority: IANR and science–science relations

As discussed in Chapter 2, how epistemic injustice is interpreted and acted upon constitutes an important dimension of the national habitus of science in Global South countries. The epistemic gap is most evident when one compares 'indigenous' or 'traditional' science with Western modern science. This section focuses on a less explored area: the role of epistemic differences in shaping which values to prioritise in emerging science. We use the example of the International Association of Neurorestoration (IANR) to demonstrate that there may still be epistemic gaps among scientists with contemporaneous Western-dominated professional training.

IANR is an international professional organisation dedicated to promoting the research and clinical application of neurorestorative therapies. It constitutes members from 40 countries, mainly China, India, Iran and Argentina. At the time of writing (spring 2021), IANR has published the second version of its Clinical Neurorestorative Therapeutic Guidelines for Spinal Cord Injury and operates two peer-reviewed English journals, with one based in China, the other in the US (www.ianr.org.cn). It also hosts the Raisman Young Scholars Awards, established in honour of the late distinguished British neuroscientists, Honorary President of the IANR, Geoffrey Raisman (Li et al., 2018).

This global entity originated in the China Spinal Cord Injury Network (ChinaSCINet), a stem cell-based neurorestorative research network of 22 Chinese centres founded in 2004 by American scientist

Wise Young. ChinaSCINet has been in the global science and media spotlight for its early exploitation of China's relaxed regulation on experimental stem cell therapies (Enserink, 2006), but also for the fact that in 2009, Young filed for US FDA approval to undertake a phase III trial in North America based on studies carried out by ChinaSCINet. At the time, it was considered a stunning feat for a China-based research network, for as Stephen Minger from King's College London commented 'To have a stem-cell trial approved by the FDA based on studies in China would be rather extraordinary' (Minger cited in Qiu, 2009: 607). To date, Young has expanded his 'China model' to set up SCINetUSA, SCINetIndia and SCINetNorway (https://keck.rutgers.edu/research-clinical-trials/clinical-trials).

In what follows, we review how Young's colleague and former post-doc, Hongyun Huang, navigated through epistemic disputes and transformed from a maverick scientist in ChinaSCINet to the founder of IANR. However, there are two important points to bear in mind. Firstly, although stem cell-based neurorestorative therapy is not devoid of ethical concerns, it is a legitimate field of research. Our focus, as with IANR's, is on how it establishes scientific credibility. In other words, in the absence of scientific certainty and prior standards, epistemic discrepancies are negotiated over what risks and benefits should be recognised and prioritised. Secondly it is important to categorically distinguish the difference between fraudulent enterprise in the name of science and the exploration of innovative therapy. Given its lucrative prospects, there are many private companies, with little or no appropriate professional background, who carry out illegal or fraudulent treatments for short-term financial gain. In fact, scientists involved in IANR, such as Wise Young, have openly criticised these ventures (Judson, 2006).

Until 2007, despite detailed disclosure of research procedures and data and an inviting attitude towards Western colleagues' site visit requests, views of ChinaSCINet and clinical trials carried out by its members were mostly sceptical if not outright negative (Judson, 2006; Qiu, J., 2007: 59) China's slack ethical oversight aside, a key point of dispute was methodological. That is, instead of the randomised controlled trials (RCTs), a gold standard of the global (emphasis on Western) life sciences, participating centres relied on self-comparisons and patients' own testimonials. Critics said that this made it difficult to rule out placebo effects, while others felt that long-term patient observation was required before any clinical conclusion could be drawn (Qiu, J., 2007).

Hongyun Huang was one of the leading scientists in ChinaSCINet. He devised therapies using nasal cells from aborted foetuses and injected them into people's spines. The hypothesis was that the regenerative nature of these olfactory cells would renew nerve cells and in turn lessen or cure diseases such as amyotrophic lateral sclerosis, Parkinson's disease and multiple sclerosis. Similar to many other experimental stem cell therapies, it was difficult to identify a standardised way to trace effectiveness (Regal, 2018). His therapy was even more unconventional in that instead of RCTs, Huang used videos, case reports and scored some patients on tests designed by the American Spinal Injury Association and the International Medical Society of Paraplegia (Enserink, 2006).

Huang was both a prudent and stubborn man. As the UK newspaper the *Guardian* reported, on the one hand, he 'promises nothing. He claims no miracle cure. He admits he cannot fully explain his results. All he knows, and all he tells his patients, is that his method often works, that the results speak for themselves' (Watts, 2004). On the other hand, he was stubborn in defending *his* ethical rationale for refusing RCTs. As most of the patients who contacted him suffered from severe illness, Huang insisted that placebo tests in this clinical context 'are unethical because they involve cutting someone open and only pretending to treat them … I wouldn't do it. Double-blind trials only harm the patient' (Huang cited in Watts, 2004).

Huang published nine papers in China-based journals, including one in English. Yet even though his work has 'received the most scientific scrutiny', his papers were nonetheless rejected by all top international journals (Cyranoski, 2005; Enserink, 2006: 161). *Nature Medicine* cited Huang's exasperation: 'Why do Westerners see it but not believe it?' (*Nature Medicine*, 2006: 262). Being a properly trained neuroscientist, naturally Huang knew that in science, 'seeing' is not 'believing'. So what really frustrated him (and peers who were conducting similar exploratory studies) was perhaps the sobering realisation that results didn't 'speak for themselves'. For a 'fact' to have a voice, it needs to be obtained, ordered and narrated in a formula dictated by (Western) doctrine. Huang's frustration was heightened by the media exposure that one of the experts who denounced his work, James Guest, plagiarised Huang's research data (Zhu and Wu, 2006). This allegedly plagiarised paper described the 'rapid partial recovery' of one of Huang's patients, an 18-year-old Japanese boy and was published in the most distinguished journal in this field, *Spinal Cord*, with Huang's name only noted in an acknowledgment (see Guest, Herrera and Qian, 2006; see also one of Guest's criticisms of Huang's method, Dobkin, Curt and Guest, 2006).

In a later publication, Huang openly questioned the implicit double-standard when ascribing validity and legitimacy to research findings (Huang 2010). But he and his colleagues associated with ChinaSCINet also realised that for their work to have a fair chance of being critically assessed by their global peers, they needed to be more proactive.

In October 2007, Huang established the IANR and defined 'neurorestoratology' as a 'sub-discipline of neuroscience that studies neural regeneration, repair and replacement of damaged components of the nervous system' (www.ianr.org.cn). While operating in parallel to ChinaSCINet, the founding of IANR effectively ushered in a much wider range of experimental regenerative medicine for one epistemic community at the margins of contemporary life science. Its 2008 a preliminary Charter emphasised an 'inevitable' 'historical' mission for like-minded scientists to come together in 'advocating science and medical ethics, promoting social justice, developing academic democracy and proposing the scientific spirit of "devotion, innovation, precision and cooperation"' (www.ianr.org.cn/xhjj).

Thus, one important departure from ChinaSCINet was that IANR was much more vocal in democratising academic thinking. For example, in Huang's introduction in the first IANR special section in the journal *Cell Transplantation*, he was firm on the point that 'the randomising double blind control study is a very important tool to assess effect in clinical trials. But it is not the only feasible tool; sometimes it even is unavailable for some treatments or studies, such as organ transplantation' (Huang, 2010: 129). In fact, Huang went on to draw similarities between patients with severe neural injury and neurone disease (which severely damages cognitive and physical capacities and may cause death) and patients with critical organ failures. He argued that just as patients' self-comparison was suited to evaluating efficacies of organ transplant treatments, it should also be considered as 'the best and simplest tool' to assess the effect of cell transplantation (Huang, 2010: 130). He then appealed to clinicians to use 'rational, reasonable and practical research methods to do study, but not follow the mechanical, doctrinal or rigid way' (Huang, 2010: 130). A year later, IANR's *Beijing Declaration*, a mission statement signed by scientists from 18 different countries, made a similar epistemic appeal that 'the importance of small functional gains that have significant effects on the quality of life' should be respected and recognised, even if the underlying mechanism remains unclear (IANR, 2009: 228). This call for more weight to be given to 'practice-based' evidence was a repeated theme in IANR discussions (Alok and Al-Zoubi, 2016).

At the heart of this emphatic call for 'academic democracy' in judging scientific validity was the fact that IANR did not see itself as a replacement or an 'alternative' to existing Western standards. They did not see practice-based evidence and RCTs as mutually exclusive but as complementary. It is more accurate to say that they expanded a shared epistemic scope with global peers. An examination of IANR's annual conference records shows that, since its founding, IANR as a group gradually incorporated RCT into its practice. For example, the focus of IANR's 2019 annual conference hosted presentations on a mixture of double-blind controlled trials and non-ramdomised trials (Chen et al., 2019). Huang himself also started incorporating RCTs into some of his studies and IANR never shied away from supporting evidence-based medicine originating in Western academia.

Soon after its establishment, IANR founded online and offline platforms to render visibility to works carried out by marginalised partitioners. In 2009 alone, the network announced three official publication platforms, which include US-based journal *Cell Transplantation* (2020 impact factor 3.477, ranked 15 out of 26 in *Cell & Tissue Engineering*). Offline, IANR's annual conference provided a venue for neuroscientists working on alternative forms of clinical interventions to share results and compare notes that would otherwise be invisible (Roseman and Chaisinthop, 2016). Since IANR's first conference in Beijing in May 2018, these annual events have travelled to a number of cities in the Global South: Amman (2011 and 2019) Bucharest (2013), Mumbai (2014), Tehran (2016) and Buenos Aires (2017). In 2018, Wise Young hosted IANR's annual conference at the University of Rutgers in New Jersey, which is, to date, the only IANR event in the West. Through the years, IANR have joined forces with the UK-based Global College of Neuroprotection and Neuroregeneration (Leng et al., 2012), the Indian Society of Regenerative Science (formerly known as the Stem Cell Society of India) and the American Society for Neural Therapy and Repair (Sharma and Sharma, 2011). In short, IANR have established an expanding international scientific community through online and offline socialisation as well as multi-channel idea exchanges.

In the most detailed and engaging investigation on Huang's clinical work to date, anthropologist Priscilla Song (2017: 180) concluded that, despite mixed results, credible efforts were made to develop alternative forms of evidence-based clinical practice. It is noteworthy that Huang and his colleagues have actively developed online forums and blogging platforms for patients to record, share and 'peer-review' each other's surgical experience and physical changes (Song, 2017: 181–96). It turned 'a singular physical event' of an individual patient into 'a co-constructed, ongoing process' of making sense of a new therapeutic approach and posts 'a poignant challenge to what counts as expertise and data' in evidence-based medicine (Song, 2017: 181–96). Thus, it seems that IANR has also taken advantage of and expanded the role of a 'civic epistemology' in the 'republic of science'.

In 2016, IANR first trialed the publication of a 'national guideline' on neurorestoratology in China. Following this, in 2017 it launched its global *Clinical Neurorestorative Therapeutic Guidelines for Spinal Cord Injury*, which covered issues such as informed consent, cell quality control, indications and contraindications for undergoing cell therapy, documentations, safety and efficacy evaluation, and refers to the US and Europe on policies regarding repeated treatments, compensation for clinical trials and publishing responsibility (Huang et al., 2018; 2019). In 2019 they updated the guidelines with a clearer ambition in offering a systematic introduction to a 'neurorestorative therapeutic standard' for clinical practice from evaluation and diagnosis, to treatments and managing complications (Huang et al., 2019: 14). There may well be disagreement about neurorestorative therapy, but Huang has become a notable leader if not an 'authority' in this expanding field.

As discussed in Chapter 2, similar bottom-up initiatives from scientists to challenge mainstream principles is not a West vs Rest antagonism, but a de-colonising effect of the increasing multiplicity of practices within contemporary science, instigated by a heightened responsiveness of the known unknowns and unknown unknowns. One example is the International Cellular Medicine Society (ICMS), an independent stem cell accreditation programme in the US which shares an Open Treatment Registry among its members. The organisation was founded by Dr Christopher J. Centeno, who developed experimental therapy for body injuries very similar to what many Chinese clinicians were doing to cure diabetes foot. That is, they isolate and process adult stem cells from a patient's bone marrow or synovial fluid and inject them back into the patient's body to treat fractures, torn tendons and other ailments, essentially injecting patients with their own stem cells (Cyranoski, 2010; Zhang, 2012: ch. 7). Through mirroring the epistemic structure of mainstream science, but following their own dictation of substance, ICMS was able to endorse medical innovation outside of the mainstream (Rosemann and Chaisinthop, 2016)

There has been 'a pluralisation of "internationally shared" or "internationally recognised" standards' (Rosemann and Chaisinthop, 2016: 114). We would like to push this argument further. The implication of such pluralisation lies not in adding 'another layer' of international governance, nor does it necessarily entail more epistemic complexity or a Kuhnian paradigm shift. As we've seen, both IARN and ICMS mirror mainstream structures; they expand rather than subvert existing epistemic logic. However, socially, they signal that we need to take a fresh view on the space called 'international'. It is arguably both liberating and daunting that 'international' or 'global' is a not space that is a given or a simple stitching of pre-existing communities, but can and has to be created. More importantly, these terms cannot be viewed in the singular either. Even within a same epistemic paradigm, there are still various 'internationals' that may collide, merge or simply be at odds with each other. The global has become a space that is simultaneously more tolerant and competitive.

In short, to be recognised is not to be submissive. It is true that, like many scientists in the Global South, for Chinese scientists, 'to be recognised' by their Western peers once meant to adopt Western norms, and follow existing guidance. Yet the subversive nature of IANR's development lies in the fact that Hongyun Huang realised that to be recognised could also mean to demonstrate and to perform.

The COVID effect? State–science relations in a globalised world

So far in this chapter we've examined internal relations that have significantly challenged and shaped the national habitus of Chinese life science. More importantly the national habitus is not constituted and informed just by particularities within national borders. Rather, as we've demonstrated, those relations, among scientists, ethicists and ordinary citizens, developed in dialogue, within an authoritarian state but with an ever increasing knowledge of global options and alternatives.

In this section, we use COVID-19 to highlight how global (re)actions are just as important as domestic ones in shaping China's national habitus. As the pandemic has affected almost every aspect of society, there are various perspectives and insights that can be considered. However, we do not intend to be exhaustive or comprehensive. Rather, in line with our overall focus, we select two domains that are most revealing for the aims of this book. One is 'vaccine diplomacy', which we will examine in Chapter 5; the other is the rivalry between Chinese national censorship and global media scepticism, which is the focus of this section.

It is safe to say that from the perspective of the government, COVID-19 was primarily an 'image crisis' before it was a public health one (Zhang and Barr, 2021). It brought the tension and mutual interdependence between the state and its scientific community into full light. While the Chinese government was among the first to make global vaccine pledges and to bet on its scientific power to reinforce an image as a responsible nation, authorities were also highly suspicious of their own scientists and clinicians. In the name of preserving a 'harmonious society' – an overriding socio-political priority in China since 2004 – officials authorised the deletion of reports of officials' early efforts to hide the severity of the outbreak. They also arrested those who tried to thwart the authorities' will by preserving reports about the outbreak or speaking out on social media – the best known case being that of Dr Wenliang Li who first publicised a SARS-like virus amongst his former medical school students in late December 2019, only to be detained by police and forced to sign a statement that he had spread false rumours and 'disturbed the social order' (Wang, Qin and Wee, 2020; Zhao, 2020). As early as April 2020, only a month after the WHO declared COVID-19 a pandemic, China established censorship rules over domestic COVID-19 research, with origin studies at the centre of ministerial-level scrutiny to curb the 'China virus' narrative (Gan, Hu and Watson, 2020).

Government censorship was deeply frowned upon by scientists and ethicists alike in China (Zhang, J. Y., 2021), as global openness was key to their professionalisation. During the first two months of the pandemic, more than 60% of the research papers were contributed by Chinese labs (Lan et al., 2020). By the time China's Ministry of Education issued the new Directive to enforce systematic censorship on COVID-related research in April 2020, Chinese scientists had already published 6.6 times more on COVID than they did on SARS in 2003 (Cai and Wang, 2020; Gan, Hu and Watson, 2020). In other words, the Chinese scientific community established its influence by being more willing and more capable of sharing its findings with the world.

Yet as COVID science became increasingly politicised, some Chinese scientists Joy was in contact with seemed to see the 'necessity' of government controls over information: 'because Western media always *xiashuo!*' said one geneticist, repeating a line the author had heard from others during previous fieldwork. *Xiashuo*, literally translated as 'talking blindly', means talking nonsense or spreading rumours. The tension is most evident from the WHO's investigation on COVID-19.

Soon after the WHO's first visit to the Wuhan Virology Institute on 3 February, a reporter from an international newspaper asked Joy whether China's censorship presented an 'insurmountable' barrier in reaching an impartial conclusion on COVID-19's origins. At the same instance, her phone displayed headlines from a news portal maintained by the Chinese Society of Biotechnology and Chinese Academy of Science. Citing the nationalist newspaper *Global Times*, the update read that Western 'smearing' of the WHO investigation was rooted in denial of the 'truth' (Biotech-China, 2021).

The international broadsheet called for impartiality in the investigation while the Chinese science platform's citing of a state-run nationalist newspaper cried out for the impartiality of the global audience. More than a year since the beginning of COVID, mutual scepticism between the Chinese scientific community and Western media seemed to be normalised into a state where both sides felt that they could predict the other's response to the origins study.

Less than a week later, the WHO dismissed the 'lab leak' theory. Media opinion that this was 'unlikely to settle the debate on virus origins' (Yang, 2021) only reinforced the mutual scepticism between the Chinese scientific community and Western media. One ethicist told Joy that she felt deflated by the Western reaction, as she believed they were fighting a losing battle: 'Normally you'd expect the investigation to be "innocent until proven guilty", but since it's a Chinese lab, the logic is "guilty until proven innocent" … These are completely two different games.'

It must be noted that at the time of writing, the investigation of the lab leak theory is still ongoing. It may take years for this discussion to settle (or to be dimmed out of public sight). We are not in a position to draw any conclusion on the validity of the claim *per se*, as we must always have an open mind about future evidence. However, what *can* be observed already is a degradation of trust relations between China and the West.

For Chinese scientists, the unrelenting scepticism exhibited by the media was particularly hurtful as many felt that their efforts in maintaining dialogue had been punished by a deep-rooted political bias. In a widely circulated interview in *Scientific American*, the Chinese virologist Shi Zhengli, known as 'Batwoman', openly shared her own suspicion of a lab leak in a March interview (Qiu, 2020). Less than a year before the pandemic, Shi and her colleagues had warned about the possibilities of a future bat-borne coronavirus epidemic in *Viruses* and *Nature Reviews Microbiology* (see Cui, Li and Shi, 2019; Fan et al., 2019). With the new mysterious pneumonia-like disease spreading in Wuhan, Shi was among the first to suspect a 'nightmare scenario' of a lab leak. She 'frantically went through her own lab's records from the past few years to check for any mishandling of experimental materials, especially during disposal and started comparing patient samples with her lab's viral genomes (Qiu, R., 2012). For accuracy, she also sent the samples to external labs for testing. Fortunately for Shi, none of the genome sequence of the patients matched the virus her lab studied. She confessed to reporters: 'that really took a load off my mind ... I had not slept a wink for days' (Qiu, R., 2012). It felt like Shi's frank confession of her worries came back to bite her.

After the investigation team left China, leading international papers, such as the *New York Times* and the *Wall Street Journal* gave damning reports, claiming 'China Refused to Hand Over Important Data' (Hernandez and Gorman, 2021; Page and Hinshaw, 2021). Such headlines were supposed to be a summary of some of the WHO expert's views. For example, Danish epidemiologist Thea Fischer was quoted by the *Wall Street Journal* that while she 'had seen no inconsistencies in the data that were made available in Wuhan', she was not given access to raw data. This was important, for as a scientist, she couldn't 'just trust what anyone tells me.' To highlight the frustrations, the newspaper also quoted Fischer's description of the exchanges between the WHO team and their Chinese counterparts, 'Sometimes emotions have run really high' (Page and Hinshaw, 2021). 'In the end', the *New York Times* wrote, 'the WHO experts sought compromise, praising the Chinese government's transparency, but pushing for more research' (Hernandez and Gorman, 2021).

Depending on one's perspective, the reports simultaneously confirmed an old (Western) conviction that there was always something to hide in Chinese science (Cheng, 2018; Yang, Y., 2021) and the old (Chinese) conviction that the West was determined not to trust China even when 'no inconsistencies' were found (Hernandez and Gorman, 2021). More importantly, despite (or because) Western commentators were criticising China 'in general', for Chinese scientists, the consequences could be very personal. For many mid-to-advanced-career scientists based in China, such generalisations often had repercussions on their own research, including their chances of publication and collaboration or seeking an overseas fellowship (Zhang, J. Y., 2021).

Western media's China-bashing became evident when WHO team leader Peter Daszak was rebuffed at the *New York Times* with a series of tweets:

> This was NOT my experience on @WHO mission. As lead of animal/ environment working group I found trust & openness w/ my China counterparts. We DID get access to critical new data throughout. We DID increase our understanding of likely spillover pathways. (original capitalisation and abbreviations) (11:27 AM 13 Feb 2021)

> It's disappointing to spend time w/ journalists explaining key findings of our exhausting month-long work in China, to see our colleagues selectively misquoted to fit a narrative that was prescribed before the work began. Shame on you @nytimes!' (2:07 PM 13 Feb 2021)

Thea Fischer also protested on Twitter:

> This was NOT my experience on the Epi-side. We DID build up a good relationship in the Chines/Int Epi-team! Allowing for heated arguments reflects a deep level of engagement in the room. Our quotes are intendedly twisted casting shadows over important scientific work. (original capitalisation and abbreviations) (2:03 PM 13 Feb 2021)

The perceived 'blindness' of Western media in COVID reporting may be seen to justify and reinforce China's nationalist narrative. The impact may be more apparent among the upcoming generation of scientists who were born in the 1990s. Dubbed 'Generation Xi' by *The Economist*, this generation not only grew up with China's steady climb in global influence, but also received the most patriotic education since Mao (Studer, 2021). This generation also represents most of China's overseas returnees. According to the latest official and commercial statistics, 86.28% of overseas Chinese students have chosen to return to China, most of whom are 'Generation Xi', seeking employment in finance and science and technology sectors (Ministry of Education China, 2020; Studer, 2021). According to Togocareer, a leading recruitment company specialising in overseas returnees, in 2020, the number of overseas Chinese seeking employment back in China increased by almost 34%, far higher than the previous two years' annual increase of around 5% (Togocareer, 2021). There is an obvious 'COVID effect' here: 'better controlled pandemic', 'more convenient life' and 'faster economic recovery from the pandemic' were the top reasons for overseas Chinese students to return. But in addition to expressed confidence in China, the survey also shows that 22.1% overseas returnees 'worry that the international climate is not favourable to overseas career development', and 17.9% decided to settle back in China because their 'resident country had adopted policies that were less friendly to Chinese' (Togocareer, 2021). There seemed to be some bitterness in the reversal of a 'brain drain' to 'brain circulation'. As the Chinese government uses high profile young returnees such as life scientists Feng Shao and Lin Wang and physicist Yuan Cao (e.g. Cingta, 2018; Overseas Intel, 2020; Translational Medicine, 2018) to propagate the idea that 'science may know no boundaries, but scientists have nationalities', the global politicisation of COVID seems to be in danger of playing into China's nationalist narratives.

National habitus, as confirmed by the COVID blame game, is not solely shaped by ideas, resources or agencies that are within the nation. Rather the structure and culture of the Chinese scientific community is shaped in dialogue with the global society. At the beginning of the twenty-first century, global scepticism arguably spurred a burgeoning Chinese life science community to adopt Western norms. Most scientists, bioethicists and policy makers we have been in contact with valued candid exchange with their global peers. But what they see as an uptick in China-bashing has riled them and made them more sympathetic to Chinese nationalist propaganda. The consequences are not limited to China. James Wilsdon and James Keeley (2007) warned that the fear and suspicion that nationalist protectionism can breed is a danger to global scientific collaboration as a whole. The stakes are now even higher, given that 49 out of the 50 fastest rising research institutions are in China (Nature Index, 2019) and that Beijing and Shanghai are the most productive city partnership in the world. In 2019, the two cities' bilateral collaborations were over 50% higher than the second most productive pair, New York and Boston (Jia, 2020): one needs to re-calculate the cost of widening distrust between China and the world.

Conclusion

Science has always been a 'politicum', a venue for wrestling for legitimacy and influence over collective actions (Linkova and

Stockelova, 2012). For the Chinese government, science lies at the heart of its struggle to rejuvenate the image of China as a powerful nation. Compared to other post-colonial regions, such as Latin America, Chinese universities are under higher pressure to demonstrate their values to techno-economic progress (Gu, 2001; Sutz, 2003). In one of the first State Council meetings on the drafting of the 2021–35 plan, Premier Li Keqiang, emphasised the need to give scientific institutions more freedom in exploring market-led innovation (Zhang, H., 2018). While the government wishes to maintain a tight grip on science publications and keep scientists in line with its harmonious narrative, it also wants to society to absorb some of the financial burdens.

But more importantly, this chapter demonstrated how science inevitably gives rise to new forms of social agency and social relations, even within an authoritarian regime in which science is often (wrongly) perceived as centrally organised. These new forms of agency aim to be recognised, that is, to be respected and included in collective deliberations. In fact, from the government's perspective, it must look like everyone is acting 'out of term'. The Chinese public may yet be given a seat at the policy table, but their 'voice', in the form of consumer rejection, has already shattered government's international pledges on measles eradication and the progress of GM research. BGI sets an example of how world-class research can be done outside of state-run institutions, with perhaps a bit more efficiency. It seems that even bioethicists, a profession that only came into the Chinese vocabulary in the 1990s, have acquired more weight than government officials in legitimising scientific research. From the Wild East episode to Huang's CRISPR experiment and IANR's global guidelines, Chinese scientists have trail-blazed an epistemic space for different forms of scientific evidence. Internationally, both Chinese ethicists and scientists are more capable of contesting rather than following Western authorities.

One needs to climb out of the epistemic boxes that have colonised our conception of how science 'should be done' to appreciate the changing relational dynamics that are steering research activities in China. We draw attention to the de-territorising nature of these new social relations in that they were developed through taking advantage of transnational exchanges. The boomerang effect is evident in both scientists' and bioethicists' strategies to strengthen their personal and/or collective professional status. Sometime, the global response may shape the national habitus in undesirable ways. The space of the 'international' has also been stretched to accommodate an emerging plurality in science. The public's ability to act on their scepticism was also related to knowing global alternatives and how science 'could be' delivered. As this chapter demonstrates, even for an authoritarian government, it may need to reconsider its relevance to and actual leverages in these new social relations. We further explore the implication of this new space for plurality in science and how governance can maintain its relevance in the next two chapters.

4

India: self-sufficiency in a globalised world

Visions of an *Atmanirbhar Bharat* or 'self-resufficient India' have guided the development of Indian sciences since its independence in 1947. In India, science has always been imagined across its multi-layered social divides as the key mechanism that would bestow the nation and its peoples the self-sufficiency it needed to determine the destiny of self and the nation therein. Each year millions of the nation's pupils sit the joint engineering examinations to become engineers, technologists, software developers. Science, for India's millions, is the ticket to social mobility. Thus, events such as the country's successful nuclear tests in 1998 or sending a Mars mission at a tenth of NASA's budget, invoke nationalist pride in the upward-mobility of 'Indian science' – to determine their own destiny in global science, and maybe, someday, lead it as well. Science for Indians is hope. Hope for the self, a better quality of life and for the nation. In the upward mobility of Indian science, the pupil of science, the Indian citizen, sees hope for their own future. In turn, science governance has always implicitly or explicitly been envisioned around enabling an *Atmanirbhar Bharat* for realising the broader purpose of nation building (Department of Science and Technology, India, 2020; Government of India, 2020).

Yet, what *Atmanirbhar* entails and how it is to be achieved has always been in response to transnational dialogue and, like the pursuit of most scientific development, hardly ever without contestation and contradiction. Rather, one may think of acquiring global visibility in contemporary science akin to opening layers of Russian Matryoshka dolls – of endless layers of power structures. In addition to dealing with hierarchies within research labs, and disparities between domestic institutions, scientists in India also need to overcome a power imbalance in influencing the global discourse. Thus, similar to what Karin Knorr Cetina demonstrated in *Epistemic Culture*, Indian scientific development at the global level relies on a handful of elites who translate information across borders, attempting to place Indian research communities onto the same page as the West. This, however, has its challenges. The uppermost echelons of India's science superstructure have access to public resources and formulate policies through the vantage points of a favoured handful of research institutions. Whereas other stakeholders, such as researchers in smaller universities or lesser hospitals, civil society groups and ordinary citizens, have limited political and financial resources to have any meaningful impact on national science policies. Thus, the voice that articulates what the national scientific agenda should embody is less a reflection of a 'universal' ideal than divisions between the *haves* and *have-nots* in India.

Consequently, within India, public views of contemporary science have become ambivalent. Before India's economic liberalisation in the 1990s, Indian science relied not only on national institutions, but also on a handful of Western-educated talented scientists with international linkages who transferred knowledge back to India through the creation of public research facilities that are critical to India's life sciences today. Yet, since the 1990s, this has undergone a shift similar to the Chinese public as communities in India have developed a similarly more reflexive attitude on the role of science in their individual and collective wellbeing.

Globally, there is also a growing ambivalence towards India. For the main part of the twentieth century, India had been quick to adopt Western governing structures, and had been a cooperative and competent provider, both in terms of natural and human resources, and in global scientific research. However, entering the twenty-first century, India, despite the government's continuous compliance with global discourse, has become a 'trouble maker'. Indeed, with examples such as farmers revoking government deals with multi-national corporations and entrepreneurial clinicians forging ahead without international consent, India appeared to lack (from a global perspective) either the competence or willingness to keep a strong hand on its researchers. Although India may not yet be *leading* global science, it is clear that scientific advancement in India has been *pulling and pushing* global science in various ways that force attention.

This chapter extends our investigation on the de-colonising imperative of science governance from the previous chapter. We continue our examination of how evolving social relations gives rise to new forms of agency by focusing on two critical events in India: the Bt crops saga and the failure of the BRAI Bill (2000–14) and the global controversy over experimental stem cell therapies (2005–18). These two cases present different and complimentary views of science in India. Research on GMOs has effectively been a 'state science' in India, whereas experimental therapy represents what we call 'science at large'. The Bt crops saga and the making of the BRAI Bill largely concerns the building of science's domestic legitimacy, while experimental stem cell therapies are emblematic of India's struggle to acquire and sustain a 'good actor' reputation beyond its shores. From a pragmatic view point, the de-colonising imperative can be simply put as this: for governance to be *effective*, it has to stay *relevant* to the subject it aims to govern. Before any authority can command science, it needs to first update its capacity to recognise and be able to speak to an increasingly diverse groups of practitioners. The two Indian cases in this chapter reveal how and where the relevance of conventional governing rationale fall short.

This chapter is structured as follows. The first section, as with the chapter on China, provides an overview of Indian life science governance. It maps out how the structure and culture of scientific practice rooted in India's colonial past have developed through successive reforms of the national science policy alongside the establishment of new regulatory branches. This provides essential information on the 'national mood' and the elitist tradition that forms the background of the two critical events we examine in depth.

The second section uses the GM saga to illuminate the national habitus of Indian science governance: the ambivalent social relation with technocracy in India. Central to India's Bt crops saga is the question 'who is "worthy" of being heard?' One striking characteristic of the Bt crops disputes was that there was no readily available categorical term to distinguish the pro- and anti-GM camps, for they were both formed by a coalition of government institutions, scientists, civil groups and industries, and both evoked a post-colonial rhetoric and the necessity for 'good science'. There was no easy way to know or to predict who must hold which view about GM crops, and positions could change over time as well. So one had to closely follow the public debate as it evolved. Conventional ways of designing and delivering regulations can easily be trapped in a self-referential 'bureaucratic amplification of credibility' which has limited ability to speak, let alone respond to diverse risk preferences.

Central to the global controversies stirred up by Indian experimental stem cell therapies discussed in the third section is the question 'who could do science?' We intentionally chose the modal verb 'could', instead of 'can', for we argue that a proper understanding of the question requires a vision that not only acknowledges those who are permitted or desirable to conduct science, but also those who are not yet but nevertheless have the capacity to do so. We draw special attention to two intriguing details of this critical event. One is how Geeta Shroff's Nutech MediWorld clinic, in particular, became emblematic of an Indian regulatory enigma to Western life science communities, while at the same time, there was also strong resistance to engaging with her to understand where the difference originated from. The other point is that, arguably, Geeta Shroff is much more subversive than Chinese scientists Huizhen Sheng, Junjiu Huang and IANR-affiliated researchers discussed earlier. For however unorthodox their scientific approaches were, those Chinese scientists were committed to seeking global peer recognitions through conventional means, such as publication. Yet, for almost two decades after the founding of Nutech MediWorld, Shroff simply 'refused' to publish her work on stem cell therapy. As she was content with remaining an outsider to institutional science, our presumed values and leverages to keep her 'in line' as a 'scientist' immediately became invalid. To some extent, this is not too different from the Bitcoin CRISPR baby venture inspired by Jiankui He that discussed in Chapter 1. Yet, as we demonstrate, these non-institutional scientists, such as Shroff, nevertheless have an instant 'universal' impact in an interconnected world. Science is thus 'at large' for it now spans outside the physical constraints of research institutions, creating challenges outside of the conventional paradigm of governance.

Finally, the fourth section concludes with an examination of India's recent national science and innovation strategy. In particular, we focus on the proposed 'one nation one subscription' initiative and its implications for the various themes discussed in this chapter.

The evolution of India's life science governance

As discussed in Chapter 2, modern life sciences was established in India through its colonial experience. Ever since, the governance of science has been a dialogue between India and the West, as well as 'a dialogue between an interrogating present and an interrogated past' (Crombie, 1990: 5)

In many ways, India's colonial experience facilitated its structural and cultural adaptation to a Western-dominated science. In the early half of the twentieth century, newly Western-educated cohorts became fellows at the Royal Society, and founded academic platforms such as Indian Journals of Science or Medicine, and indigenous professional societies such as the Bombay's Grant Medical College Society (Gorman, 1988: 293). The Western-educated intelligentsia readily adopted practices such as peer-review processes and professional societies, while the so-called 'unscientific' systems of indigenous medical standards (based on Ayurvedic and Unani traditions) were structurally weakened by the increasing trickle-down adoption of British medical standards (Bala, 2012). The adoption of the UK's General Medical Council's licensing system formed the basis of administrative control over indigenous knowledge systems (Jeffery, 1979) and remains entrenched today across the government in India, including in the life sciences.

British influence is similarly visible today in the elitist culture amongst the Indian scientific community where a few Western-trained professional elites sit at the top with a plethora of non-privileged and local-trained researchers at the bottom (Lawrence, 1985). For instance, when the state funded Indian Research Fund Association (IRFA) was set up in 1911 to promote medical research in India, it was intended to be almost entirely staffed by recruits from the Indian Medical Service (IMS), which had served the colonial military and bureaucracy since 1764. The Medical Research Council, created two years later in 1913, and the IRFA shared many characteristics – both were autonomous, intended for knowledge-led 'pure' science research, funded through public and private means, and run by an Advisory Board, which in IRFA's case was entirely European. After independence, IRFA was renamed the Indian Council of Medical Research (ICMR) in 1949, and used for promoting research in universities. Since 1953, with the Rockefeller Foundation's assistance, several research units in existing colleges have been set up, in line with the foundation's interest in promoting Western education and scholarly linkages with the US. By the early 1950s, the rise of US global hegemony on the back of its scientific achievements became India's, along with many other nations, shining example to follow a similarly technoscientific path to prosperity, if not global leadership.

Yet, within this context, the technocratic science governance that emerged under Jawaharlal Nehru, the country's first Prime Minister after independence, closely resembled the Western science focus dominated by the cold war calculus. The bulk of India's science funding was devoted to defense-related big science projects and big dam projects reflecting the successes of US science such as its much-feted Hoover Dam project. 'Nehruvian science', was expected to 'alone solve the problems of hunger and poverty, insanitation and illiteracy, of superstition and deadening custom and tradition, of vast resources running to waste, of a rich country inhabited by starving people' (*New York Times*, 1946).

National identity became entangled with science development envisioned by Nehru and a small group of knowledge elites (Krishna, 1991; Prakash, 1999). Post-independence science governance became dominated by mission-oriented funding initiatives (Krishna, 2001). What was needed at the time, after independence, but never quite emerged was an assessment of societal needs. Small agriculture-based projects to increase agricultural productivity or medical research to reduce the incidence of communicable diseases were ignored. This eventually led to national humiliation at a scale that still sets public opinion against Nehru today. Indeed, Nehru's neglect of and failure to invest in agricultural productivity research, despite repeated US warnings of failing crop growth since at least 1959 (e.g. 1959 Ford report), saw the country go begging throughout the 1960s to the US and other nations for grains, and to its own citizens for 'dinnerless days' (*TIME*, 1959, 1960, 1965). Furthermore, the long-term impact of Nehruvian science policy's skew towards specific mission-oriented strands of 'strategic' importance (Sharma, 1976), namely defence, space and atomic energy, were substantial, commandeering over 55% of the nation's gross expenditure on research and development well into 2013 and beyond (Krishna, 2013).

The establishment of its Department of Biotechnology (DBT) in 1986, as one of the few nations with an agency dedicated to progressing agricultural and health biotechnologies back then, provided further affirmation of India's progress and impetus for its global aspirations. According to the DBT 'The remarkable march of India into the world of biosciences and technological advances began in 1986. ... That decision [to set up the DBT] has made India one of the first countries to have a separate department for this stream of science and technology' (DBT, 2017). The vision of the DBT's first secretary, Dr Ramachandran, on the DBT's launch was that at the pace which biological sciences were growing globally, 'unless we [India] leap[s] forward, there is no way of catching up with the rest of the world' (DBT, 2017). His vision reflected not only India's global aspirations in life sciences but also the need to be an early mover in the field. Yet, knowledge elitism persisted. There were considerable overlaps of expert membership across various government departments, committees and subcommittees, which had the effect of concentrating funding and power in the hands of a few. The situation was worse with new scientific areas as there were a small number of specialised experts to start with. A 1981 report on 'Health for all: An alternative strategy' commissioned jointly by the Indian Council of Social Science Research and the ICMR stated that the 'health sector is exotic, top-down, elite-oriented, urban-based, centralized and bureaucratic with over-emphasis on curative care' (Subramanian, 1981).

Likewise, in the area of bioethical considerations for human subjects of biomedical research, the country viewed itself as an early mover, having published the Policy Statement on Ethical Considerations involved in Research on Human Subjects in 1980 – within just a year of the publication of the globally acclaimed Belmont Report on the Ethical Principles and Guidelines for the Protection of Human Subjects of Research by the United States Department of Health and Human Services in 1979 (HHS, 1979). Similarly in 1994, India became one of the first few nations to ratify the United

Nations Convention on Biological Diversity when it entered into force on 29 December 1993. Subsequently, it had established its own Biological Diversity Act as early as 2002 – a year before the Cartegena Protocol on Biosafety entered into force on 11 September 2003.

Beyond the public sector research and development landscape, private industry, particularly in pharmaceuticals, flourished, bolstered by a unique set of social conditions that had evolved favourably since the 1960s and *in spite* of the systemic challenges just discussed. Until 1956, 99% of the drug manufacturing industry of India was in the hands of ten foreign multinationals who followed the imperial model of importation–re-exportation to keep medicine costs in India at their highest globally (Greene, 2007: 2). Indigenous pharmaceutical production was modest. Western medicines and formulations used by the colonial state were raw materials that were imported from India and re-exported as finished drugs at a huge mark-up (Chaudhuri, 1984). This allowed a handful of British pharmacies to profit hand-somely from supplying the European communities (Greene, 2007: 2). However, a licensing policy-change in 1956, making it mandatory for multinational companies to have manufacturing in India, lowered entry barriers and allowed indigenous manufactures to enter the market along with multinationals. By 1972, the suspension of 'product patents' from the India Patents Act (1970) compelled the departure of foreign multinationals. By the time India launched its economic liberalisation policies in 1991 and opened its market to the world, the pharmaceutical industry had built up sufficient capacity and economies of scale to export generics to the world. Today, India's pharmaceutical sector supplies 'over 50% of the global demand for various vaccines, 40% of the generic demand for US and 25% of all medicines for UK' and boasts 'the second largest share of phar-maceutical and biotech workforce in the world' (IBEF, 2021).

Following economic liberalisation throughout the 1990s, the national mood in India underwent a gradual change, from being distinctly pro-development in the early 2000s, to dual pro-development and pro-economic progress towards the latter half of the decade. The pro-development agenda of the early years started with an administrative change from the ruling Bharatiya Janata Party (BJP) to Congress Party-led United Progressive Alliance (UPA) coalition in May 2004 'despite the excellent performance of the economy

... [and] toward greater state activism in economic affairs' (Nayar, 2005). For the 2004 election, the then ruling BJP's *India Shining* election platform was intended to showcase the country's above par performance across key economic indicators and its growing presence in the global arena both economically and militarily (given India's rise to a nuclear power in 1998 under the BJP). Yet in the elections, the *India Shining* manifesto backfired against the UPA's populist pro-poor policy proposals for the common man, with 'future promise of free electricity and a job for each poor family – even if through make-work programs' (Nayar, 2005). Subsequently, Congress' consecutive electoral victory five years later in 2009 entrenched the pro-poor 'focus on job creation, [and] infrastructure investment, besides expansion of social sector programs' (Sharma, 2011). Yet, to achieve its objectives, the UPA administration's policy strategy followed 'a more inclusive governance approach ... designed to pursue the liberalising agenda ... a dual approach of seeking to appease the powerful middle-class constituency while appealing to the economic majority ... a deeper strategic purpose of achieving centrism and a broad-based social coalition' (Hasan, 2006).

In this endeavour, the UPA administration's 2004 mandate to deepen 'relations with the United States' (Kronstadt, 2004), coupled with the US President Obama's visit to India in November 2010 pledging to 'promote trade, boost exports', saw a move towards entrenching centrist pro-business goals (Sharma, 2011). Importantly, India was keen to forge international collaborations as its economy had been struggling since the global recession and the nation was moving towards one of the most substantial fiscal deficits at 7% of GDP (Sharma, 2011). In this context, the changing position statements for 'biotechnology' in the government's Five Year Plans (FYP) evidences a shift away from purely developmental policy trajectories to international trade-oriented strategies as in the excerpts from the FYPs in Table 4.1.

In 2002, the expectations from investment in biotechnology research and development had been its contribution to inward developmental goals: its 'impact on food, nutrition, health, environment and livelihood security' (PCI, 2002). By 2007, these expectations from investment in biotechnology had shifted to its contribution to outward-looking goals for making India 'globally competitive' (PCI, 2007). By 2012, this had been ramped up to attaining 'global

Table 4.1 India's last three FYPs before the dissolution of the National Planning Committee

Five Year Plans (FYPs)	Drafting party	Key points
Tenth FYP (2002–7)	Formulated by the BJP administration's economic focus before UPA administration came to power in May 2004	10.107 India is well poised to embark upon biotechnology-based national development. **The underlying assumption of the policy framework is that the development in the field of biotechnology will have the greatest impact on food, nutrition, health, environment and livelihood security** … Long-term support would be provided for basic biology research in areas related … stem cell research, etc. by providing the necessary infrastructure and instrumentation facilities … facilitated by instituting new academia-industry and private-public partnerships (PCI 2002) (added emphasis)
Eleventh FYP (2007–12)	Formulated by the newly elected Congress-led UPA 'pro-poor' administration in power from 2004 to 2009	8.54 The approach pf the Department of Biotechnology (DBT) has been to create tools and technologies that **address the problems of the largest section of the society** and provide them with biotech products and services at affordable prices. The ultimate objective is to make India **globally competitive** in the emerging bio-economy … Developing a strong biotechnology industry and technology diffusion capacity is critical for fulfilling this objective. 8.55 The priority areas … include … stem cell research and regenerative medicine. **International cooperation activities match national needs and the above priorities will also be accelerated** (PCI, 2007) (added emphasis)

Table 4.1 India's last three FYPs before the dissolution of the National
Planning Committee (Continued)

Five Year Plans (FYPs)	Drafting party	Key points
Twelfth FYP (2012–17)	Formulated by re-elected Congress 'pro-poor' administration in power from 2009 to 2014	8.39 … to support the overall plan of the Indian Science, Technology and Innovation sector towards **global leadership.** … Several new initiatives of DBT for the Twelfth Plan period have been prepared taking into account **national needs and likely impact.** 8.56 The overall strategy for DBT … is to advance biotechnology as strategic area by taking India's strength in foundational sciences to globally competitive levels and expanding the application of biotechnologies for overall growth of bio-economy **within the framework of inclusive development** (PCI, 2011) (added emphasis)

leadership' but within the parallel objective of achieving it 'within the framework of inclusive development' (PCI, 2011). Viewed through the lens of John Kingdon's (1984: 153–6) national mood of the polity, these parallel objectives of global competitiveness and inclusive development frame the ruling Congress administration's political need to appeal to the diverse moods of its diverse electoral base, which were primarily (a) the private sector and the upper middle classes, increasingly interested in participating in the benefits of global resource flows, and (b) the 'common man' or the middle classes and the poor, interested in the benefits of inward job creation. In turn, this differentiated approach of appeasing the electorate along class lines is mirrored in the DBT's and the ICMR's position in backing similarly differentiated and parallel aims of 'greater private participation 'mirroring the 12th FYP's 'inclusive development' goals (*Business Standard*, 2008; PCI, 2011). The 12th FYPs 'global leadership' goals are reflected in supporting international collaborations and focuses on key strategic areas such as stem cell research and its clinical applications (*Business Standard*, 2008).

The balancing act between global collaboration and indigenous capacity building is perhaps best emphasised by K. Kasturirangan, eminent space scientist and member of India's Planning Commission in 2013, when he says that,

> science and technology held the key for India's progress in future and there are many new challenges stemming from scientific fields such as nuclear energy, genetic engineering, synthetic biology, nano technology and stem cell research … There is a need to devise a proactive process to be adopted for assessing public attitudes on the one side and the level of risk acceptable in a social context of evolving policy. (*The Hindu*, 2013)

Biotechnology and the lifesciences were thus not only inextricably aligned with the political-national mood of the nation but also reflective of national aspirations and identity for the broader purpose of nation-building. In turn, this focus of India's national policies and guidelines on 'appeas[ing] communities' (Kingdon, 1984: 157–8), both inside and outside of India by striking balances between competing narratives on what good science and good science governance are, becomes visible in both of the critical events this chapter discusses.

The BRAI Bill and social ambivalence towards technocracy

While the only approved GM crop in India to date is Bt cotton, the public debate in India on GM is perhaps one of the liveliest anywhere. This is perhaps not surprising, for one key difference between India and other emerging scientific powers, such as China and Brazil, is that agriculture remains a crucial part of its national economy. Whereas agriculture accounted for 4.4% and 7.7% of the national GDP of Brazil and China respectively in 2020, for India, the figure was 19.9% (Kapil, 2021; Statista.com). The difference is more striking if one compares it to developed countries, such as the US and Japan, where agriculture contributes less than 1% to national GDP. In addition to the pressure of feeding 1.4 billion people, farming is the primary source of livelihood for about 58% of India's population. More importantly, according to the latest census of the Indian Ministry of Agriculture and Farmers Welfare (2019), 86% of farming households have 'small' (1–2 hectares) or 'marginal' (less than 1 hectare) landholdings.

GM debates have a significant impact on India's science governance. For example, the first time an Indian state government took a multinational company to court was in 2006. Andhra Pradesh's Department of Agriculture sued the American agrochemical giant Monsanto for charging abnormally high prices to farmers for Bt cotton seeds; other states later joined the fight to protect their local regulatory autonomy (Raina, 2013). It was also the partnership with Monsanto over the development of Bt brinjal that dragged India's major seed company, Mahyco, into a series of public controversies, including being India's first commercial entity to be accused of biopiracy (Sood, 2012).

The climax to this GM debate in India came in 2010 when a bill to set up a Biotechnology Regulatory Authority of India (BRAI) was proposed by the DBT. The BRAI's purpose (according to the bill) was to create a national-level 'single-window fast-track' system in which a five-member committee (three full-time, two part-time) was proposed to replace states and other national level authorities in granting approvals to GM crops. It further proposed that not only would committee members be subject to an 'oath of secrecy' (section 9.2) but that some of the information submitted by biotechnology developers could be classified as 'confidential commercial information'. Critics argued that both of these clauses would not only effectively remove information from public view but were unconstitutional as they went against the purview of India's *Right to Information* Act. Quickly nicknamed by the public as the 'Monsanto Protection and Promotion Bill', the protest against the BRAI Bill eventually coalesced into the 'Monsanto Quit India' movement, a revocation of Gandhi's 1942 'Quit India Movement' which called on mass civil disobedience to demand an end to British rule in India. Protestors drew comparison between the threat to India's collective autonomy as 'Back then, it was the East India Company and now we have

"Eat India Companies!'" (Sood, 2013). Thus, it is unsurprising that, from the global perspective, the GM debate in India is often framed as a fight between the indigenous masses and corporate-capitalist agriculture.

Yet, in this chapter, we want to highlight another under-explored theme of the GM saga to illuminate the national habitus of Indian science governance: that of the ambivalent social relations of technocracy in India as exhibited by the public outcry surrounding the proposal of the BRAI Bill. Since the late 1980s, the Indian government has perhaps been one of the most proactive governments globally in initiating GM crop regulations. Soon after its promulgation of the New Policy on Seed Development in 1988, which offered pro-market regulations, the inter-ministerial Genetic Engineering Approval Committee (GEAC) was founded in 1989 under the Rules of the Environmental Protection Act. GEAC's remit included approval of the open field trials and commercial release of GMOs (www.geacindia.gov.in). In the 1990s, India supported the development of both domestic and transnational private ventures in the research and commercialisation of seeds, which has been a symbol for its economic liberalisation and globalisation commitment (Scoones, 2006). The fear of being 'overtaken' in science by China, along with domestic needs, had prompted the DBT to invest in 19 GM crop projects in the 1990s, all but one of which were transnational collaborative projects using transgenes developed from OECD countries or multinational firms (Indira, Bhagavan and Virgin, 2005). While the GM research in India is a joint force of private and public venture with transnational support, given the central role the Indian government play in creating the policy, and institutional and funding environment in organising the research, it is essentially a 'state science' (Herring, 2015).

By the mid-1990s, civil society in India had started to voice concerns over the exploitation of farmers, the environment and other social issues connected to GMOs alongside similar protests the world over (Kumar, 2016; Shiva, 1997). In the UK, the GMO debate had erupted in February 1996 after the Sainsbury's and Safeway supermarket chains started selling GM tomato puree. Prince Charles weighed in against GM crops saying that the reality 'was enough to send a chill down the spine' (BBC, 1996). In 1998, the European

Parliament placed a moratorium on further GM crop approvals even though it contravened WTO rules (*Financial Times*, 2006). In the US, 7 organisations launched a million-dollar nationwide campaign to pressure Congress and the Food and Drug Administration to 'force premarket safety testing and labelling [of GMOs]' (Roosevelt, 2000). Countries in the Global South were also becoming hostile to GMOs. By the year 2000, the backlash against GMOs was 'exploding 'from every corner, with "government ministries, farmers, unions, consumer organisations, and environmental groups clash[ing] over whether to allow commercial planting of GMOs" across the Global South' (Schurman and Munro, 2013: x). Remarkably these protests were not necessarily confined to their national borders nor were they constrained to the short term. For example, as Schurman and Munro (2013: xxvi) noted, 'although activism in the United States had a negligible impact on the domestic policy environment, US activists played a crucial role in creating the Cartegena Protocol on Biosafety which became a major resource for activists in poor countries to pressure governments'. Indeed, in India, these global narratives were slowly trickled down to grassroots level by activists and transnational non-governmental organisations (NGOs) such as Greenpeace India. The 2003 Cartegena Protocol eventually become a touchstone of the activism against GM crops in India and against the BRAI Bill, as many demanded the need for a protection-focused 'Biosafety Regulatory Authority of India' instead of the proposed biotechnology-promotion centric BRAI (Coalition for a GM-Free India, 2011).

Thus, at one level, the rationales behind the GM debate in India echoed those in other parts of the world. The pro-GM camp often cited increased productivity, reduced pesticide use, food security, better food quality, less labour, plus other associated economic benefits, as reasons to widely adopt GM crops, while opponents cited food safety, cross-pollination, pest resistance, risking seed (and thus) an exploitation of sovereignty (Indira, Bhagavan and Virgin, 2005; Scoones, 2006; Stone and Glover, 2011).

At another level, both proponents and opponents evoked a post-colonial rhetoric and the necessity of 'good science'. More importantly, as demonstrated below, the 'indigenous masses and the corporate-capitalist' framing can be misleading as the two camps cannot easily be demarcated into East vs West, or the masses vs science. Rather, at least domestically, both sides were formed by a coalition of government institutions, scientists, the industry and civil societies. A key conflict thus lies in *the organisation of how GM is governed*. To frame it more explicitly: who gets to govern and whose views are taken into account? To understand this in the lead up to the proposal of the BRAI Bill and why it pushed the controversy to a climax, one must start with the Bt brinjal and Bt cotton rows that preceded it.

The Bt brinjal moratorium and the BRAI Bill leak

The contentions embedded in the BRAI Bill began in 2000 with the Government of India's approval of controlled field trials for Bt brinjal. Bt cultivars genetically modify the germplasm by inserting the *Cry1Ac* gene, an insecticidal protein derived from a common soil bacterium, the *Bacillus thuringiensis* or *Bt*. This allows the Bt cultivar to develop in-built protection against pests (MoEF-GEAC, 2009). In turn, this technological modification is typically pitched as a key tool for enhancing agricultural adaptability to climate change, feeding rapidly expanding global populations and increasing farmers' income. Additionally, there were practical concerns that made Bt brinjal, in theory, socially desirable in India.

In India, brinjal is mainly grown by small growers throughout the year, due to the vegetable's adaptability to various climate conditions. However, given brinjal's high susceptibility to the *fruit and shoot borer*, which reduces marketable yields by as much as 60–70%, repeated insecticide use becomes necessary but costly at around Indian Rupees 5952 per acre (about 80 USD) – a substantive outlay for India's subsistence farmers (MoEF-GEAC, 2009). One survey conducted in Bangladesh revealed that farmers had to spray a single brinjal crop 84 times in a 6–7 month cycle (BARI, 1995 in MoEF-GEAC, 2009). Aside from costs, this degree of intensive insecticidal spraying carries health risks – both for farmers from repeated exposure (Kaur, 2008) and for consumers from higher residual traces in edibles (Kumari et al., 2004; Mancini et al., 2005; MoEF-GEAC, 2009). Environmental degradation through groundwater and soil contamination is also an added concern. At the same time, the not insubstantial acreage devoted to brinjal cultivation, accounting for 8.14% (0.512

million ha) of the land under vegetable cultivation in India and contributing 9% (8.4 million metric tonnes in 2007) of the country's total vegetable production, meant that the gains from improved yield had the potential to substantively increase food resources while improving farmer's incomes (MoEF-GEAC, 2009).

Thus, the brinjal was an easy choice for launching GM food crops in India, and its development was actively encouraged by the DBT in the Ministry of Science and Technology and the Ministry of Environment and Forest (MoEF; now the Ministry of Environment, Forest and and Climate Change (MoEF&CC)). Maharashtra Hybrid Seeds Co., better known as Mahyco, started developing the transgenic Bt brinjal as early as 2000. Mahyco is one of India's leading seed research and marketisation companies and has recently expanded operations to other South Asian countries, such as Vietnam, Indonesia, the Philippines and Bangladesh. Notably, 26% of Mahyco was owned by the US-based Monsanto. In 2004, India's review committee on genetic manipulation (RCGM) under the DBT, allowed five varieties of Bt brinjal to be tested for biosafety. These tests were conducted across 11 multicentre but small-scale limited trials in the research farms of the state agricultural universities under the Indian Council of Agricultural Research. Two years later, in 2006, the RCGM concluded the efficacy and safety of Bt brinjal before recommending it for approval to the GEAC as 'safe to the environment, nontoxic, nonallergenic and ha[ving] potential to benefit the farmers' (MoEF-GEAC, 2009: 2). Subsequently, the GEAC placed a summary of RCGM's biosafety report on its website for public consultation (http://envfor.nic.in). At the same time, GEAC convened the first Expert Committee on Bt brinjal (EC-I) for reviewing evidence pertaining to the Mahyco application. In July 2007, EC-I submitted its own recommendations to the GEAC informed by Mahyco's own biosafety data submission and stakeholder submissions. A few months later, on 8 August 2007, the GEAC granted Mahyco permission to conduct large-scale trials of seven Bt brinjal cultivars for two seasons 'under the direct supervision of Director, Indian Institute of Vegetable Research (IIVR), Varanasi' (MoEF-GEAC, 2009: 2). A further three-member sub-committee under the National Centre for Agricultural Economics and Policy Research was assigned the task of conducting an 'ex ante assessment of socioeconomic benefits' of Bt brinjal cultivation (MoEF-GEAC, 2009: 24). Additionally, a second expert committee (EC-II) was also constituted in 2009 to examine the data from these trials. Eventually on 31 August 2009, the GEAC approved Mahyco's application for large-scale trials and seed production, concluding that 'the data requirements for safety assessment of GM crops in India are comparable to the internationally accepted norms in different countries and by international agencies and therefore no additional studies need to be prescribed for safety assessment' (MoEF-GEAC, 2009: 29). However, the GEAC deferred the final decision on the open-air field trials and seed production to its parent – the MoEF – who, in turn, placed the issue for nationwide consultation from 13 January to 6 February of 2010. Just two days after the consultation ended, Jairam Ramesh, the then Union Minister of State for MoEF placed an indefinite moratorium on the release of Bt brinjal on 9 February 'till such time independent scientific studies establish, to the satisfaction of both the public and professionals, the safety of the product from the point of view of its long-term impact on human health and environment, including the rich genetic wealth existing in brinjal in our country' (*The Hindu*, 2010). However, it was soon public knowledge that Indian government ministries were split on this issue. Weeks later, it was leaked that the DBT had already started to draft the BRAI Bill to not only provide fast approval for GMOs but also to centralise (and restrict) all biotechnology approval decisions to one apex national committee. The draft was leaked in March 2010 and provoked an immediate public outcry around its constitutionality and ethical challenges (Mehta, 2013). However, to fully comprehend the relations, tensions and emotion that developed around Bt brinjal it is necessary to contextualise it within the co-evolving social negations and contestations around the governance of Bt cotton.

The Bt cotton 'fad' and its opponents

Bt cotton has been approved for commercial cultivation in India since 2002 when India became the sixteenth country to approve genetically engineered cotton for environmental release. Thus, by the time Ramesh placed a moratorium on Bt crops in 2010, the breadth of the environmental, ethical, legal and social implications of open-air cultivation of a GM crop was already under discussion in the country. Like Bt brinjal, the inserted Cry gene in Bt cotton was expected to provide in-built pest control, and to increase yields, and had been developed by Mahyco Monsanto Biotech (MMB), a fifty-fifty joint venture between Mahyco and Monsanto for primarily developing, producing and sublicensing Bt cotton seeds in India. After the Government of India cleared MMB's Bollgard-I variety of Bt cotton seed for commercial cultivation in 2002, MMB licensed it to seed companies in India for sale to farmers. An updated Bollgard-II was also cleared in 2006.

In the beginning, Bt cotton was met with favourable curiosity. In the key cotton-growing region of Warangal in Andhra Pradesh, for example, the introduction of this hybrid seed was characterised as the 'Bt cotton craze', or in anthropologist Gleen Davis Stone's words 'a fad' (Stone, 2007: 69). Between 2003 and 2005, the market share of Bt seeds climbed from 1% to 62%. In some villages, Bt cotton took over the whole crop (Stone, 2007: 68). In many ways, Bt cotton ratified the government's push on GM technology, turning India from a major importer of cotton to a major exporter. Little more than a decade after its commercial cultivation, Indian media heralded that 'India has beaten China to become the largest producer of cotton in the world' (*India Today*, 2015).

However, there is always another side to the same technology, and Bt cotton was not a harbinger of prosperity for everyone. Accusations of MMB's oppressive monopoly practices and exorbitant pricing, which led to the exploitation of small farmers and farmer suicides, had started even before cultivation had begun on Bt cotton in 2002 (Lal and Malaviya, 2013). Simultaneously, various 'technical issues' of the Bollgard variety also came to light: for instance, its longer 180 to 200-day growing period compared to the quicker 150 to 160-day cycle of traditional varieties, which stretched farmer's investment capacity; and its negative impact on yields of traditional (non-transgenic) cotton varieties. (Bhagwat, 2012). There were also biosafety concerns, as many Indian firms also produce and market 'bio-insecticides' containing Bt spores to kill caterpillars (Herring, 2015). As a result, civil groups demanded clearer biosafety research.

NGOs were also highly vocal in pressing the central government in Delhi to address biosafety and environmental concerns (Coalition for a GM-Free India, 2011). By 2007, farmers were mobilised to protest against Bt cotton and MMB's monopoly practices such as artificially creating market shortages to sell at exorbitant black-market prices, and substandard seed quality. These initiatives had some effects. For example, Vidarbha is a key cotton-growing region of India. In 2012, when the state of Maharashtra banned MMB seed sales, Kishore Tiwari, president of the non-governmental farmers' movement Vidarbha Jan Andolan Samiti, which had agitated for and been instrumental in the ban, accused 'Bt Cotton [of] play[ing] a key role in accelerating the farmer suicide saga in Vidarbha since June 2005' (Pawar, 2012). Domestic industry was on the side of the farmers too. The seed industry lobbying group, the National SeedAssociation of India spearheaded calls for a reduction in Bt royalty fees. This eventually led to a 74% nationwide cut in 2016, which slashed costs from Indian Rupees 163 per bag to 43 (Bera and Sen, 2016).

Criticisms were met with little tolerance from the pro-GM camp. It immediately triggered worry, then condemnation, from research groups and biotechnology firms. After the central government slashed Monsanto's royalty fees in 2016, Shivendra Bajaj, Executive Director of the Association of Biotechnology Led Enterprises Agriculture Group, accused the move as being against free-market principles and warned the government of its 'short-term populist measures' over 'supporting innovation in the long term' (Bera and Sen, 2016). Many scientists shared the worry that if India did not give this forward-looking biotechnology sufficient support, it would once again fall behind the rest of the world. In short, as leading scientists protested: 'it is bad for the country' and 'actually affect[s] the indigenous effort' (Jayaraman, 2010). This view was embedded in the government's push for GM technology, as described in the next section.

The BRAI Bill and the 'bureaucratic amplification of credibility'

As successive Indian governments had been unwavering supporters of GM science, the aforementioned Jairam Ramesh's 'indefinite moratorium' in early 2010 shocked the GM world. In the eyes of GM supporters, what qualifications did a Minister of Environment and Forests have to make a judgement about *science*? If one talks to practitioners in India, it is common to hear of the need for sciences to be governed by a 'competent body'. However, such 'competence' is often narrowly defined as coming from expert bodies constituted by a small circle of like-minded academic scions. Sometimes, such as in the making and discussion of the BRAI Bill, there is a paradoxical phenomenon, where the more rounds of the debate, the narrower the range of contributions.

To paraphrase Roger E. Kasperson and his colleagues' (1988: 181) definition of the 'social amplification of risk', we argue that in India the national governance of GM resembles a 'bureaucratic amplification of credibility', which denotes the phenomenon by which professional legitimacy, institutional ambition, group think and elite manoeuvres shape the public perception of credibility, thereby amplifying the weight of a particular claim. In what follows, we explain why the DBT's intention of promoting Indian socio-economic welfare with good science went askew by demonstrating how various 'competent bodies' related to generating or validating the evidence-basis of the bill effectively generated a bureaucratic amplification of credibility.

At surface level, there is a 'chain-of-accountabilities' that formed such bureaucratic amplification. Each 'technical' competent body shored up the 'technical' rigour of the other. For example, the GEAC's final recommendation to the MoEF in 2009 was based on the recommendation of the RCGM and the Expert Committee-I, which in turn was informed by the Expert Committee-II. However, a more profound reason for amplification is that the 'eligibility' to sit on committees and contribute to policy discussions was often restricted to a small circle of technical experts. For example, it is not unusual for an elite scientist to sit on different government committees, resulting in the same opinion being channeled through different institutional voices.

The initial risk assessment of Bt brinjal was almost solely done by a group of close-knit scientists and government officials. Voices from civil society, social scientists or farmer groups were largely absent. International scientific standards were equated with social legitimacy. For example, in the EC-II's 2009 report, it was noted that international scientific standards were equated with social legitimacy, as examplified by the conclusions drawn by the previously cited 2009 GEAC expert committee report (MoEF-GEAC, 2009: 29). Abiding by scientific standards doesn't guarantee ethical research, nor does it guarantee social legitimacy (Hurlbut, 2018). However, the all-scientist EC-II committee not only validated the prior decision of EC-I merely on the basis of technical compliance, but also declared that 'no additional studies' were needed. In effect, it foreclosed dissent. To be sure, the report did include 'issues raised by NGOs, national and international groups' (MoEF-GEAC, 2009: 54). However, each concern was translated into a technical problem that was solved by technical experts following the technical standards set by international authorities such as the US Food and Drug Administration, and Codex Alimentarius (MoEF-GEAC, 2009: 54–66).

To fully comprehend how hand-picked 'competent bodies' create a bureaucratic amplification of credibility, one also needs to take into account the 'mutual accreditations' that take place within India on who is 'worthy' of being heard. For example, dissenting views, such as those of biologist Professor Pushpa M. Bhargava, a vocal critic of GM crops and the founder of India's Centre for Cellular and Molecular Biology at Hyderabad, were dismissed as an insensible obstruction to development. As the EC-II report stated: 'Raising the bar of the regulatory process as recommended by Dr P.M. Bhargava based on hypothetical concerns and apprehensions would be highly detrimental for research and development in the area of agricultural biotechnology especially for public sector institutions' (MoEF-GEAC, 2009: 61). In the aftermath of the moratorium, the MoEF Minister Jairman Ramesh commissioned six Indian academies of science to produce a rigorous review on the desirability of commercialising Bt brinjal. The Inter-Academy Report was released in September 2010 and concluded that Bt brinjal was safe for human consumption with negligible environmental effects. The 25-page report ended with two quotes:

> GM technology, coupled with important developments in other areas, should be used to increase the production of main food staples, improve the efficiency of production, reduce the environmental impact of agriculture, and provide access to food for small-scale farmers. (Royal Society of London, the US National Academy of Sciences, the Brazilian Academy of Sciences, the Chinese Academy of Sciences, the Indian National Science Academy, the Mexican Academy of Sciences, and the Third World Academy of Sciences, In *Transgenic Plants and World*

Agriculture (2000), Document made available by the Indian National Science Academy, New Delhi)

The affluent nations can afford to adopt elitist positions and pay more for food produced by the so-called natural methods; the 1 billion chronically poor and hungry people of this world cannot. New technology will be their salvation, freeing them from obsolete, low-yielding, and more costly production technology. (Dr Norman E. Borlaug (Nobel Prize Laureate for Peace 1970), Inter-Academy Report, 2010: 24)

We draw attention to these quotes, one of which comes from 6 major academies in the world, the other from the American agronomist who championed the Green Revolution, as they illuminate the mentality underlying this report: a technoscience imagination coupled with the insecurity of a post-colonial nation. 60 years on, the GM debate still echoed strongly with the same 'science-society contract' set out by Pandit Jawaharlal Nehru (Gupta, 2011). It was later found that part of the report plagiarised materials from a US-based lobbying group, and was considered a 'sham' (Jishnu, 2011). The report's rather amateurish 'investigation' carried out by elite scientists was perhaps not surprising. If the legitimacy and validity of a view has long been reified and amplified through various bureaucratic exercises, why would one bother to look for more evidence?

The side effect of the bureaucratic amplification of credibility is that it also legitimises ignorance under the pretence of efficiency. As we noted before, the BRAI Bill not only proposed to centralise national decision power over GMOs into the hands of five technocrats, but also the right to remove certain information from public view to protect decision making under 'commercial confidentiality'. In addition, the BRAI Bill replaced the constitutionally enshrined jurisdiction of states over their own agricultural welfare to an ambiguous advisory role (BRAI Bill, 2013, ch. VIII: 17–18). Critics pointed out this was unlawful as it violated the 73rd and 74th amendments of the Indian Constitution on the 'decentralisation of power' (*The Hindu Businessline*, 2013).

The BRAI Bill was presented to parliament in August 2013. During the parliamentary debate of the bill, a survey of 440 scientific research studies on the environmental and health impact of GM crops were released. In addition to civil society protests, hundreds of scientists wrote a letter to the Prime Minister, demanding all open-air field trials of GM crops be paused in India until further safety infrastructure was established (Todhunter, 2014).

The bill eventually failed to pass. Yet, in the eyes of many, the harm was perhaps already done: public confidence in both science and the government was shaken; GM scientists felt a sense of lost autonomy; and the government and industry were disillusioned by the delay in reaping socio-economic returns from their huge investment. When the Bt backlash erupted against the central government's and the pro-GM policies, similar to other countries, the Indian government turned to its epistemic community, the expert groups, to handle public concern (Javali, 2004). However, the problem was that the nature of the expertise they relied on remained the same (Salter and Jones, 2002). The BRAI Bill, drafted to tackle growing public scepticism of GM crops, seemed to have only aggravated the problem. The policy deliberations that led up to the bill were underpinned by a bureaucratic amplification of credibility. The mutual support and confirmation among a small circle of scientific scions had effectively blinded them from engaging with social realities.

Reflections on the BRAI Bill and social ambivalence towards technocracy

The Bt crop saga and the BRAI Bill was an iconic event in re-tuning science–society relations and in shaking up science governance in India. It is not difficult to see how the story could be cast as small farmers versus the seed industry, or ordinary citizens against centralised technocratic rules. It shows how the elite epistemic community remains a key broker in Indian science policies. However, it is also an example that demonstrates how civil epistemology is still significant in shaping the social uptake of a technology and how science–society relations remain contingent (Gupta, 2011; Jasanoff, 2015).

Yet, the examination of this social ambivalence towards technocracy doesn't end there. Its implications are more profound. For despite strong opposition to GM crops and the BRAI, it is important to note how similar the camps of proponents and sceptics actually were. Both sides were formed by a coalition of government institutions, industry groups, scientists and civil groups, and were informed by a myriad of global discourses. In this sense, it was simultaneously a domestic discussion and a conversation with the world. Both sides also evoked the necessity for social advancement and the importance of sound science. Let's not forget the initial commercialisation of Bt seeds were characterised as the Bt cotton fad. The difference was that both sides had competing visions of how social advancement should be measured and by whose account. They also had different approaches on the selection and interpretations of scientific evidence. As Ronald Herring (2015: 180) insightfully argued, 'science itself has no normative claims about risk preference ... there is no authoritative guide to how much evidence is enough evidence, how cautious is cautious enough. Nor can science conclusively rule out consequences unexpected from current settled knowledge.' Having scientifically produced data is not the same as acquiring social legitimacy. Contrary to the assumptions in the 6 Indian academies' Inter-Academy Report, the availability of facts and standards opens up a conversation rather than bringing it to a close. The bureaucratic amplification of credibility blinded knowledge elites from seeing that the legitimisation of science had been decolonised away from the ivory tower of (transnational) academia and the desk of visionary leaders.

The bureaucratic amplification of credibility is by no means a uniquely Indian phenomenon but India, given its traditional narrow definition of scientific expertise and governing structures, has a magnifying glass effect on its consequences. The social ambivalence towards technocracy exhibited by the Bt crops saga was directed neither towards science nor towards bureaucracy per se. Rather it was a campaign for the 'varieties of modernity' rather than an either/or solution (Zhang, 2015). Ambivalence is not indecisiveness but highlights the evolving nature between technological possibilities and the social evaluation of choices (Jasanoff, 2015). In other words, Indian societies at different levels have a shared desire to march towards modernity instead of being roped in by a singular vision. After all, the aim of scientific development is not to close down our life options but to open them up.

Science at large: experimental stem cell therapies and its discontents

Science is at large. This is not only because of the worldwide recurrence of the citizen science movements such as the emergence of do-it-yourself biology or a young generation of researchers shaped by high-profile global events, such as the International Genetically Engineered Machine Competition (Irwin, 2018; Landrain et al., 2013;). Science is at large, for, as we've seen from the opening scandal of this book, even within conventional institutions, such as hospitals and universities, and among conventional professions, such as clinicians, geneticists or embryologists, there has been an increasing blurring of boundaries over who has the capacity to do what and where they have the authority to do it.

Years before Jiankui He, experimental stem cell therapies had already warned us that cutting-edge scientific explorations have exceeded existing regulatory regimes. Both China and India had acquired some notoriety for experimental stem cell therapies. We have discussed the IANR case and how it challenged the scientific epistemology in the previous chapter. Building on that, we want to examine experimental stem cell therapies in the Indian context and provide another perspective on how it urges a rethink of our epistemology of governance. More specifically, we start our discussion with one particular clinician: Geeta Shroff.

Saheli first heard of Geeta Shroff in 2013. She had just joined the UK Economic and Social Research Council-funded project 'Rising Powers' as a research assistant. The project's focus was on state strategies of governance in global biomedical innovation, especially in regard to China and India. During one workshop planning meeting, her colleagues discussed inviting the New Delhi-based clinician, Geeta Shroff, to speak. Shroff, through her founding of the controversial Nutech MediWorld clinic, in New Delhi, India, had acquired something of a 'celebrity' status in the world of unproven stem cell therapies. As her colleagues went on to discuss the details of the conference organisation and who else to invite to speak, Saheli was still caught in bewilderment: 'Geeta Shroff, who is she?' For someone who regularly checked Indian news, this was not a name of any significance in Indian domestic discussions.

It's not that Saheli didn't know experimental stem cell therapies had been carried out in India. She knew of a few trials conducted in India's most prestigious public medical facilities. For example, Dr V. S. Sangwan, a renowned ophthalmologist, along with ocular biochemist Professor Dorairajan Balasubramanian, Director of Research at the L.V. Prasad Eye Institute in Hyderabad, had been administering stem cell therapies to patients with eye disorders since at least 2001 (Sangwan et al., 2005; Vemuganti, Sangwan and Rao, 2007). She had also read about Dr Venugopal, Director of the All-India Institute of Medical Sciences, who had been testing a cardiologic stem cell theory for end-stage heart patients since at least 2003 (Mudur, 2005; *The Hindu*, 2005). What puzzled Saheli at the time was why her British colleagues had opted, at least from the perspective of many Indians, for a much less 'notable' doctor in a private clinic.

Who is Geeta Shroff? According to the biography published on NuTech MediWorld's official website, Shroff was a homegrown Indian gynaecologist who had obtained her training from the University of Delhi. Through 8 years of practice in public hospitals in Delhi, she became a qualified IVF practitioner and started working on stem cell therapies at the end of the 1990s. Although at the time of writing (spring 2021), Shroff claims to have 'over 50 publications in international peer reviewed medical and scientific journals' (www.nutechmediworld.in), it was not until 2015 that the first publication of her NuTech practice came out (Shroff and Barthakur, 2015). At the time of Saheli's meeting, Shroff still 'refused' to publish her results (Laurance, 2011). Her rise to stem cell 'stardom', or more precisely, notoriety, was through events such as her 2005 press conference, attended by India's then Health Secretary Prasanna Hota, Minister Ajit Jogi, and the Prime Minister's wife Gursharan Kaur, in which she claimed to have treated 100 patients with stem cell therapies (PTI, 2005). However, without any publications, it was hard to verify her claim. So her practice was considered by most Western bioethicists as borderline medical quackery. As leading stem cell expert at the King's College London, Professor Stephen Minger, commented, her claim was 'highly implausible and frankly downright dangerous' (Ramesh, 2005).

A few weeks after the initial workshop planning meeting, Saheli and her colleagues reconvened again. Of the conference invitations they had sent out, one politician was bemused by the idea of extending a platform at a leading Russel Group university in London to Geeta Shroff and had simply responded with a question: 'Is this a con or a joke?'

How did a 'quack' at a private clinic, rather than esteemed professors at renowned institutions became the face of India's experimental stem cell therapy? Why was she not discussed much at home? How was it that while she posed an enigma to Western ethical governance of regenerative medicine, inviting her to share her views was (at least to some) an incomprehensible and downright ludicrous idea? Our discussion addresses these questions. More importantly, through our analysis, we want to point to what these phenomena reveal about India's governing rationale, the 'epistemic disobedience' of developing countries and, as we set out in the Preface, the new 'universality' of science.

How did experimental stem cell therapy became a policy problem in India?

India has always prided itself on having foresight in developing the life sciences. The birth of the world's second test-tube baby in Kolkata on 3 October 1978 was widely celebrated as a demonstration that Indian scientists could compete on the world stage (Bharadwaj, 2016: 55). This was followed by further research investment in the 1980s and 1990s, including collaborations with highly reputable Euro-American laboratories. In stem cell research, the successful cloning of Dolly the sheep by Scotland's embryologist Ian Wilmut in 1997 followed by US-based Thomson's successful culture of human embryonic stem cells (hESC) a year later in 1998, started a race of global science discovery. For India, the race in a new research realm was an opportunity to gain a foothold in the Western-dominated global life sciences, especially as, on 9 August 2001, the US President George W. Bush introduced a ban on federal funding for research on newly created human embryonic stem cell lines. By 14 September 2001, the Parliament of India had organised a brainstorming session at the Institute of Pathology, New Delhi, which had concluded that 'stem cell research and its clinical applications should be promoted in the country' (Parliament of India, 2005a: section VIII). A committee of experts for drafting guidelines was almost immediately instituted by the ICMR to formulate guidelines.

Despite the political recognition of stem cell therapy as a new *scientific* area, for Indian regulators, it was not seen as a new *governing* topic. For example, in 2002, the first draft guidelines for stem cell research had been placed on the ICMR's website for public consultation. The 2002 draft guidelines substantially mirrored the text of India's *Ethical Guidelines for Biomedical Research on Human*

Subjects published earlier in 2000 – which itself was a revised version of the original *Policy Statement on Ethical Considerations Involved in Research on Human Subjects* published in 1980 by the ICMR. In fact, up until the 2002 draft guidelines were updated and republished in the 2006 *National Guidelines on Stem Cell Research and Therapy*, public and policy discussions on stem cell research were scant until the Shroff story broke in mid-2005 and caught the attention of the Indian Parliamentary Standing Committee on Health and Family Welfare (hereafter referred as Parliamentary Committee). On 21 December 2005, a Parliamentary Committee discussion on Shroff's experimental stem cell therapies was held; an archival record of the proceedings noted that the

> Committee's attention has also been drawn by media reports about several clinics promising stem cell cure for diseases from muscular dystrophy, neuromuscular disorder, stroke, liver cirrhosis, diabetes, etc. in Delhi and other cities in the country. When asked about the authenticity of such claims, the Director-General, ICMR informed the Committee that very few Centres have approached the Council for approval. NuTech Mediworld, New Delhi, which had submitted incomplete papers, have been asked by the Council to submit the detailed procedure and protocol including the source of stem cell used for therapy. The applicant, however, failed to provide the same even after several reminders. (Parliament of India, 2005b: section 8.4 Recommendations/Observations)

It is safe to say that experimental stem cell therapy became a 'problem' in India through a boomerang effect of the global media, while a request to 'table' (whereby proceedings are archived for accountability purposes) the question of research on stem cells in May was dismissed by the Upper House of the Parliament. Seven month later, Shroff's news pushed the the Parliamentary Committee's subsequent recommendations to be tabled at both the Upper and Lower Houses. In contrast, in India, the media coverage of Shroff or unproven stem cell therapies in general was limited.

India undertook a series of initiatives to respond to the global gaze. The 2007 National Guidelines for Stem Cell Research and Therapy (hereafter referred as *2007–Guidelines)* was drafted in 2007. Indian authorities recognised that it was not domestic policy reform per se but how such reforms were *recognised* by key internationally esteemed scientists that could help Indian stem cell research shrug off the image cast by Shroff's clinic. One of these efforts was India's ICMR working along with the British Deputy High Commissioner in Bangalore in co-organising the week-long UK-India Round Table on Stem Cell Research and Policy Guidelines in 2008. The event was attended by leading scientists from both the Indian and UK sides, and the additional presence of India's national top officials and industry leaders exhibited the importance and political significance India attached to this meeting (*Business Standard*, 2008; Kumar, 2005; Mehra, 2008). The Round Table on Guidelines was expected to 'explore opportunities for collaboration and commercialisation of stem cell technology' with a focus on inwards investments to the UK. Nevertheless, the week-long event started on the first day with India-UK negotiations of acceptable national guidelines and the subsequent issuance of a 'Consensus Statement' where 'panelists … call[ed] for ensuring standards in collecting, processing and administering stem cells used in trials and therapy' (*Deccan Herald*, 2008).

Two months later, when the *2007–Guidelines* was jointly issued in April 2008 by the ICMR and the DBT, it contained the first extensive guidance directly impacting downstream commercialisation of stem cell therapies. What's interesting is that the *2007–Guidelines* copied verbatim many of the UK–India Roundtable's statement separating internationally acceptable Indian research (conducted by researchers in public institutions) from the unacceptable practices of experimental stem cell therapies such as in Shroff's clinic.

Indian authorities were also keen to bring their statements in line with global norms. For example, when the Indian government updated its 2007 national guidelines in 2013, it had dropped the words 'and therapy' from the title (ICMR-DBT, 2013). According to the Foreword in the *2013–Guidelines*, the omission of 'therapy'

> emphasize[d] the fact that stem cells are still not a part of standard of care; hence there can be no guidelines for therapy until efficacy is proven. These guidelines are intended to cover only stem cell research, both basic and translational, and not therapy. It has been made clear in these Guidelines that any stem cell use in patients, other than that for hematopoietic stem cell reconstitution for approved indications, is investigational at present. (ICMR-DBT, 2013: Foreword)

While closely observing international rhetoric helped India to integrate its stem cell research into the global norms and helped to develop a shared vision within its domestic research community, this did not mean that India had passively imported Western regulatory frameworks. In fact, as the next section shows, there is an epistemic gap between British and Indian stakeholders on what good governance and what good science should look like.

Who could do science?

One of the questions underlying many governing challenges of our time is: who could do science? A proper answer to this question requires not only a renewed delineation of the key characteristics that constitute the identity of 'who', but, as explained in this section, also involves a re-consideration of what 'science' means.

This question is important as it forms the root of a major divergence in governance, especially in areas of emerging subjects where scientific knowledge remains limited. One example is on the issue of stem cell therapy regulation. In the aforementioned UK–India Roundtable of 2008, while the delegates from two nations came to an agreement on guiding principles, there remains a key difference on how these principles are best observed. A contrast between the views of two attendees of the Roundtable helps to demonstrate our point. Prof Colin McGuckin, of the Newcastle Centre for Cord Blood, summarised the British perspective on governing stem cell therapy: 'We believe in the UK that not everyone should be allowed to do stem cell research and not everyone given a licence. Patient safety should not be compromised … stem cells are not the whole answer … They can't cure everything. We must not peddle false hopes of cure' (*Deccan Herald*, 2008). For the UK, a good scientist is at the heart of good science. Thus, 'who' gets to do science is a legitimate focal point for regulators. The UK was one of the first in the world to devise a governing structure that focuses not simply on the technology per se, but on guarding the entry of practitioners. Embryo research took off in the UK in the 1970s, following the pioneering research of English scientists Steptoe and Edwards into in-vitro fertilisation (IVF) in 1978. Despite considerable public and political debates over related ethics, legal and social implications, a narrative around the 'benefits' promoted by eminent stem cell scientists and the press led to a 'strongly positive image of embryo research' by the end of the 1990s (Mulkay, 1994: 210). Yet to address public concerns over the prospect of 'genetic control' and the call to 'insulate societies against the excesses of scientists', the UK parliament passed the Human Fertilisation and Embryology Act in 1990, which led to the founding of the Human Fertilisation and Embryology Authority (HFEA), an 'executive, non-departmental public body, the first statutory body of its type in the world' (Campbell-Savours, in Bloomfield and Vurdubakis, 1995: 538; HFEA, 2004: 2). HFEA covers all privately and publicly funded hESC research and therapy across national and sub-national levels and oversees the use of human embryos, gametes and stem cells in research in the UK through a licensing system. Licences are granted only if the researchers and proposed research aims meet HFEA's standards. Framed this way, gatekeepers such as HFEA are seen as a good way for life sciences governance in the UK to strike a balance between supporting scientific innovation while minimising abusive technological exploitations (Lovell-Badge, 2008).

In contrast, one Indian expert present at the same meeting, Dr Jotwani from the ICMR, described the take-away message from that day's discussion as:

> In India, the new stem cell policy [*2007–Guidelines*] … has put in place a mechanism for greater private participation and compliance. Review and monitoring of stem cell research has been decentralised at the institutional level. (*Business Standard*, 2008)

> Permissive research are disease-specific and are being addressed locally at institutional levels. Restricted and prohibited level proposals are to be referred to the central committee and this addresses broader issues like cloning, cell-based engineering and works on reproduction. (*Deccan Herald*, 2008)

For India, good judgement is at the heart of good science. More specifically, it is the case-by-case contextualised decision making by regulators and scientists at the local level. The role of good governance is not to restrict who gets to do science, but to provide general guidelines, and to reflect the need for 'greater private participation and compliance' and (at least in theory) enable local institutions to decide. In fact, while the *2007–Guidelines* strongly echoed UK norms and were promulgated by two government departments, neither the

ICMR nor the DBT had jurisdiction over clinicians. Instead, clinicians are self-governed by the Medical Council of India, much like the General Medical Council does in the UK. Such de-centralisation was also seen in the subsequent *2013–Guidelines*, which considered the best way to manage rapid advance was to increase the capacity of self-governance among practitioners. For instance, the *2013– Guideline*s note that 'It is the responsibility of the researcher and the Institutional Review Committees to understand the principles of these guidelines and keep abreast with the existing regulations in the country' (ICMR-DBT, 2013: 2). Similar to previous guidelines, the *2013–Guidelines* also lacks clarity on some of the key words, which led to an interpretive ambiguity, which is open to subjective criteria-setting by individual researchers or institutions (Datta Burton, 2018; Tiwari and Raman, 2014: 416).

In policy making, ambiguity is a fact and a virtue that 'permits conflicts to be diffused because opposing parties [favouring opportunity or concerns] can attach their own interpretation to the decision outcome' (Zahariadis, 2003: 168). This suggests that *ambiguity* would have been a desirable mechanism for striking a balance between acknowledging the need to manage risk and the need to maximise benefits. What is interesting is that similar room for discretionary enforcement and fragmented local decision making can also be found in China's life sciences. It shifts the responsibility and associated socio-political risk of carrying out research in a grey area from the national ministries to local institutions, or even to individuals (Datta Burton, 2018). Although such practices are often criticised by scientists and ethicists for lack of policy consistency essential to a stable research context, they are also seen as an instrumental tool to establish some control in an area while maximising the speed and potential of science advancement (Zhang, 2012a; 2017). In short, India and China have different epistemic paradigms from the UK for what good governance should focus on, what it prioritises and how it is best carried out. We would argue that such differences in governing outlook are not so much 'cultural' but positional. Here, perhaps a reference to the field position we discussed in Chapter 2 may be helpful. That is, for previously marginalised players such as China and India, having the chance to keep up with or develop a lead in a new research area carries a significant weight in improving their positions, whereas for players who already occupy a central position in global science, a routinised orderly development is preferable. As the *New York Times* noted stem cell research represented a '"new pot of gold" for Indian science and businesses' (Mishra, 2005), India's choice of governing outlook was a reflection of the opportunity cost of precluding individuals from contributing to this field.

In addition to different top-down regulatory visions of 'who' can do science, we also want to draw attention to divergences in bottom-up perspectives on who actually has the capacity to take advantage of new technological possibilities. We refer back to the Geeta Shroff case. Medical researchers called her a 'dangerous quack', while improved patients called their treatment a 'miracle'. Shroff herself corrected them 'it's not a miracle, it's science' (Blakely, 2009). Is it science? One of the key controversies over the Shroff case was that up until 2015, she did not publish any of her treatment data or submit it for peer-review, 'a cornerstone of good science' (Blakely, 2009; Foster and Fleming, 2007). This not only made her claims about her clinical achievements unverifiable, but her refusal to discuss how her stem cells are purified or explain how the cells functioned in the body encouraged suggestions that it could be merely a placebo effect. Others considered her a rogue. Seen this way, the UK politician's response to Saheli's colleague made sense: how can any reputable academic organisation invite a rogue for a keynote? Is it a con or a joke?

As such, peer-reviewed publication is a reoccurring issue in global controversies on avant-guard science and is a benchmark by which to credit individuals. This is also an area where Asian researchers, particularly those based in China and in India have repeatedly failed to meet the standard. As we've seen in the previous chapters, from Xigu Chen's 2001 claim, to Jiankui He in 2018, not having peer-reviewed publications seems to be a shared trait among 'rogues'. On the other hand, Huizhen Sheng, Junjiu Huang and IANR-affiliated researchers all struggled to get their data recognised in Western journals. To some extent, they all failed, as they eventually had to opt for home-based English-language peer-reviewed journals, or in the case of IANR, establish their own journals so as to communicate freely with global peers. We did not find statistical data to support the 'Western bias' some of our interviewees have expressed, nor is this point important to our purpose here. It is sufficient to recognise that peer-reviewed journals represent a particular way of scientific communication among practitioners and, more importantly, create an important validation channel to credit claims for those who identify themselves as *professional scientists*.

Geeta Shroff posed an ironic problem to the scientific world. She simply 'refused' to publish. Of course there may be various reasons and not having sufficient data at the time could be one of them. However, the point is, she didn't seem to care too much either! Shroff was more interested in applying for patents and protecting commercial secrets. Did Shroff view herself as a scientist? It may depend on one's definition. However, it is safe to say that she did not view herself as a scientist in the tradition of the other Chinese cases we've discussed. She is a business woman who happened to have substantial clinical skills and resources.

She may be accused of not following good practice by the scientific community, but if she does not identify herself as part of this community, to what extent does she need the validation of peer-reviewed journals? Furthermore, what would be the incentives for her to spend the time publishing? In other words, it seems that we need to rethink the validity and even relevance of a key governing tool.

One may argue that Shroff would need validation from the scientific community to gain the trust and credibility of her patients. The UK's *Guardian* newspaper once listed the credentials hanging on Shroff's office wall as follows: 'Indian medical diplomas, training certificates from Asian research institutes, and a picture of her with India's Prime Minister Manmohan Singh, who is a friend of the family' (Ramesh, 2005). Given the fact that Shroff charged as much as £30,000 for a single course of treatment, a lack of scientific justification to patients was highlighted by many as irresponsible and unethical. One of her American patients who regained the ability to walk reportedly said, 'I know I haven't been injected with a placebo or apple juice.' Yet that didn't matter to him, what was invaluable to him was the fact that he was on his feet again (Blakely, 2009). In a book chapter that explains her treatment, Shroff justifies the moral imperative by highlighting that her selection of patients were only those with few or no treatment alternatives remaining and that she focused on small clinical improvements rather than outright cures or increased survival (Shroff, 2018).

It would be belittling to simply see this as a case of exploitating desperate patients. A more unsettling point here is how the usual assumptions of sources of credibility and trust did not work. Shroff's clinic, and many clinics like hers, are difficult to clamp down because they have operated outside of the conventional governance paradigm.

Another point that needs to be raised is that while Shroff attributed her treatment to 'science', the role of stem cells in her clinic features more as a technology rather than as part of a research project. That is, her prime aim was not to *generate* knowledge but to *apply* knowledge, to tinker with it, so as to address a practical problem (e.g. a patient's illness). Of course through her treatments she may acquire new learning, but that, for her, was only a 'side effect'. Does she have an obligation to share it with the scientific world? Such obligation could be argued and possibly enforced, yet the point we want to highlight here is that, there was and is no such governing structure in place.

The real contention underlying experimental stem cell therapy is not the risks and benefits themselves, but the question 'who could do science?' Who is eligible and best placed to take up the task and judge these risks and benefits? Gatekeeping, both by government and by peer pressure, is a good idea. Yet how realistic is it for us to control a new area of research by controlling who does it? Shroff may not be, or care to be, a stem cell professor, but the problem is that she has the means to do the things scientists do, and without effective oversight, possibly more. Furthermore, depending on one's perspective, Shroff may not be doing (proper) science, but her experimental treatment, just as Jiankui He's CRISPR baby, has nevertheless influenced science proper. Such gaps in governance need to be addressed. This is because, for any governance mechanism to be effective, it must incentivise compliance and not drive practice underground. It must be able to speak, not just to the imaginary role of traditional academics, but to the new communities and new identities who are actively taking part in the production and application of knowledge.

Science is at large, not only in the sense that it has spanned outside the physical constraints of research institutions, but also in the sense that it has been extended to various niche social contexts that are outside of conventional governance. While it is easy to see that Shroff failed to comply with various good practices, it is difficult to fully digest the fact that she and her 'science' were operating in the same technical grammar but in a different social paradigm. That is, she relied on the same language of contemporary stem cells and followed a similar technical logic to apply existing knowledge, but her source of credibility, sense of reward, and clientele relations were very different. More importantly, she had established a niche to accommodate this new social paradigm. This is not different from many other private ventures that are vying for or have already initiated projects of 'biohacking'. One of the thorny questions Shroff posed is that, while governing tools and norms (e.g. peer-reviewed publication) play a central role in validating and legitimising one's behaviour, now it seems that the relevance of such tools and logic needs to be (re)validated. As such, communication platforms should be and must be extended to accommodate these 'mavericks', just as the Second International Summit on Genome Editing organising committee decided to give He the platform to explain his views in Hong Kong. Extending the platform does not mean we are in agreement, or even that an agreement will be reached, but it is essential to keep ourselves engaged with the real-time dynamics of science.

Who should we blame?

Experimental stem cell therapies raised several serious ethical issues: absence of consistent quality control, unknown and potentially lethal side-effects, mis-information and exploration of desperate patients, etc. Private clinics that offer unproven experimental stem cell therapies have been particularly demonised by the media and by mainstream academics (Braude, Minger and Warwick, 2005; Datta, 2018b; Giles, 2006; Qiu, 2009).

Blame could work as an effective tool to warn and discipline individuals and institutions in line with global norms. For example, as we have demonstrated in previous chapters, the worry of their international 'image', or the worry of being blamed by other nations spurred both China's and India's promulgation of stem cell guidelines. However, an under-explored topic is how blame is framed and on whom it is placed also has a latent effect on conformity, cooperation and contention in global science governance. More importantly, as scientific communities in different countries watch, contemporaneously, differentiated media and political takes on similar or comparable research, it may shake the perceived legitimacy and fairness of the accusation.

One problem with blame is that it tends to reinforce prejudice rather than helping to get closer to solving the issue. For example, Geeta Shroff effectively became the image of 'Indian' experimental stem cell therapies and her rogue reputation has been extended to India as a rogue destination (Novella, 2010; Sipp, 2009). For Indian stem cell scientists, this is a strange fixation of Western media, as they would perhaps consider someone like Dr V. S. Sangwan, whose stem cell therapy, the *Simple Limbal Epithelial Transplantation*, which has become a mainstream surgical procedure for eye-related degenerative conditions, as more representative of Indian stem cell research (Shanbhag et al., 2019). One may also ask why Ms Reema Sandhu, who has multiple sclerosis, and paid £70,000, sourced partly through GoFundMe, to the private HCA London Bridge Hospital for experimental stem cell therapies, was heralded as 'reap[ing] remarkable benefits of medical revolution', while similar practices in India were framed as dangerous and exploitative (Waters, 2019). Or why cases such as the Autism Regenerative Centre in London, which offers £9,500 stem cell treatment of autism to children over the age of two, and the existence of more than 70 private providers in the UK who offer sham stem-cell treatments for joint pain, have not attracted more intensive media or academic investigation (Curwen, 2020). Such sentiment echoes with what we described in Chapter 3 on Chinese scientists' response to China-bashing during COVID. The risk of unbalanced criticism is that it will not incentivise cooperation, but only generate greater divide between the Global South and Global North. To some extent, it has brought another antagonistic layer to the question 'who could do science'.

Another problem with blame is that it can blind us from seeing how science should and could be responsibly developed. For example, in their perceptive article in *Science*, Olle Lindvall and Insoo Hyun (2009) warned that stem cell tourism-bashing may close off the route of medical innovation necessary for stem cell therapies. Drawing on the Belmont Report, they differentiated the aim of 'clinical research' as producing scientifically generalisable results and 'medical innovation' as patient care (Lindvall and Hyun, 2009: 1664). In practice, clinical research relies on the epistemological privileged 'objective' scientific evidence such as those generated by randomised clinical trials, whereas medical innovation draws on experiential evidence which has been mislabeled as 'subjective' or 'anecdotal' (Datta, 2018; Goldenberg, 2006; Lord Saatchi, 2014). Similar to many other surgical techniques that were developed previously, stem cell therapies may require a combination of scientific routes to establish safety and efficacy. Furthermore, Lindvall and Hyun also highlighted an ethical consideration often missed:

> From many patients' point of view, consenting to medically innovative care may be preferable to enrolling in a clinical trial, especially where patient care is decidedly not the purpose of the trial – expanding knowledge is. Patients with precious little time might not care much about expanding knowledge; what they care about is getting better and surviving. Demonizing stem cell tourism will never squelch this vital instinct. Acceptable channels must be made available to seriously ill patients. (Lindvall and Hyun, 2009: 1664)

For Lindvall and Hyun, instead of using blaming and shaming to try to curb experimental stem cell therapies outside of selected institutions and/or selected elite scientists, or to educate and discipline patients' desperation, a more productive avenue of governance would be to recognise such needs and incorporate medical innovation by articulating 'acceptable conditions under which "unproven" stem cell therapies for specific diseases may be attempted, as medical innovation, in patients outside of clinical trials' (Lindvall and Hyun, 2009: 1664–5). After all, one cannot exert influence over what has been cast out. Effective incorporation of the actors and practices is the first step to establish good governance.

A reflection on science at large

It may be helpful to rewind time and have a look at how the Geeta Shroff story unravelled. India's experimental stem cell therapies first attracted wide public attention in February 2005. The highly respected *Times of India* published an exclusive story, announcing that the nation's most revered doctor, Dr P. Venugopal at the All India Institute of Medical Sciences (AIIMS), marked a global first in pioneering stem cell medicine through an injection method (Kauli, 2005). This 'path-breaking study' was carried out between February 2003 and January 2005. Venugopal's team recruited 35 cardiac patients whose conditions were beyond bypass surgery. They were given stem cell treatment and monitored at 6, 12 and 18-month intervals (Kauli, 2005). The article detailed that, 'the statistics speak for themselves. After 6 months, 56% of the affected (the dead muscle) area injected with these cells had shown improvement. After eighteen months, this went up to 64%.' Also announced was the opening of 'a national stem cell centre at AIIMS, which will coordinate the research and its applications', and that this achievement placed Indian stem cells 'right at the top of the world's medicine map' (Kauli, 2005).

Yet, statistics didn't 'speak for themselves'. If they did, then they were interpreted as articulating quite a different story by the listeners, both inside and outside of India. A few weeks later, on 17 March, the news was reproduced in *Nature* with the title 'Indian regulations fail to monitor growing stem-cell use in clinics' (Jayaraman, 2005b). The then Head of Basic Medical Sciences at the ICMR was shocked for learning this news, and said 'We are only a block away from AIIMS and we did not know this was happening there, … If the nation's premier medical institute did not ask our permission for such therapy, how can we blame private clinics for what they do?' (Jayaraman, 2005b: 259).

The fact that Indian national regulators can't keep up with research conduct 'only a block away' validated and aggravated Western scepticisms. In reality, inability to govern the rogue clinician 'a block away' occurs more than one might think and isn't restricted to the Global South. For example, US-based Dr Centeno and colleagues (2016) run around 34 clinics in the US alone, offering experimental stem cell therapy for the knee, and its StemCellArts branch at Chevy Chase, Maryland was a mere 20-minute drive away from the US Food and Drug Administration's headquarters in Rockville, Maryland. Going back to Shroff, further ethical concerns about her practice were discussed in the May Editorial of the *British Medical Journal* written by leading stem cell scientists (Braude, Minger and Warwick, 2005). In November the same year, the *Guardian* interviewed Geeta Shroff in her 20-bed clinic Nutech Mediworld and reported in an article with the sub-headline juxtaposing two points: 'West urges curb on Indian clinic's untested treatment' and 'controversial stem cell work gets patient backing' (Ramesh, 2005).

After the *Guardian*'s reportage two things happened. First, Shroff's clinic and the ethics of clinicians offering experimental stem cell therapies went viral in the British media (e.g. Foster and Fleming, 2007; SKY News, 2006, 2007; Watts, 2006). The Nutech Mediworld captured, or rather, reconfirmed the (Western) public imagination of how India endeavours to lead on cutting-edge science. Second, expert opinion on the issue pressured the UK regulator to relax regulations around reproductive technologies to hasten domestic therapeutic discovery while pressuring foreign jurisdictions like India and China to tighten regulations (Bharadwaj and Glasner, 2009; Braude, Minger and Warwick, 2005).

Both India and China constitute the geography of blame in the field of regenerative medicine. In some ways, the international reception and subsequent political characterisation of India over experimental stem cell therapy echoed the disputes China experienced on its hybrid embryo studies discussed in Chapter 3. Both countries, despite their differences in political systems, adopted a permissive and de-centralised approach to regulate emerging science. Just as Chinese bioethicists were shocked to learn of Xigu Chen's experiment, so were officials at the ICMR about clinical trials a block away. This chapter has discussed how the subaltern status of the two countries contributed to different epistemic lenses on the relation between science and national-level regulation. Both countries also had similar experiences of how domestic celebration of break-throughs have backfired and become part of their scandalous image internationally.

A few years later, evidence for the use of hESCs, and its ethical justifications for providing healthcare to terminally ill patients at Shroff's Nutech MediWorld, was submitted in writing and accepted by the UK's House of Lords Science and Technology Committee, during their consultation on regenerative medicine (UK-House of Lords, 2012). This also reminds us of the case discussed previously on Chinese scientist Huizhen Sheng's experience of being consulted by the UK government years after her research was banned in China.

There are, of course, differences between the two countries and the individuals concerned as well. For one thing, both Chen and Sheng worked at state-owned institutions with public funding, whereas

Shroff's clinic was a private venture. For another, India did not categorically ban stem cell therapies the same way China did with hybrid embryo research.

However, there seems to be some general patterns on how the West responds to cutting-edge research in the East, and how the two sides approach science and regulation differently. Through our empirical engagement with both sides of the discussions, we feel both sides have something to learn. For example, despite their (rhetorical) conformity with global discourse, India and China are epistemically disobedient. Underlying a chronic 'enforcement problem' was a socio-political rationale that projects emerging sciences and their governance with different opportunity costs and thus with different goals and priorities. Their permissive regulations and case-by-case discretions may give them a chance to race ahead, although permissiveness does not simply mean relegating responsibilities to individual institutions. Perhaps both countries should reconsider what role the government should play and what responsibilities should be taken up at the national level. The West, or rather the world together, has something to reconsider as well.

Science has become at large. Conventional governing tools, such as peer-review validations and licensing systems, which are used for gatekeeping, also seem to struggle. We need to reorient our governing paradigm from the question 'who *should* do science?' to 'who *could* do science?' The significant shift lies in the fact that the scope of agencies that contribute to or tinker with science, including cutting-edge science, is no longer something that could be designed or effectively constrained through a top-down mandate.

For governance to be *effective*, it has to stay *relevant* to the subject it aims to govern. Such relevance is built, not on what the regulators want, but on what the practitioners want. It necessarily needs to be able to speak to their aspirations and desires before it can establish any influence. To some extent, one could argue that different ways of thinking about governance in India and in China are what give rise to cases such as Shroff, but only to some extent. Shroff captured Western attention perhaps partly because she presented an enigma about who could 'afford' to be defiant to conventional scientific communities – communities she didn't align herself with but whom she impacted nonetheless. As global science has become ever more 'cosmopolitanised' (Zhang, 2012a), the legitimacy and authority of the global governance of science is becoming ever more dependent on its perceived fairness and inclusivity of diverse groups of practitioners.

Science for the masses? Indian's outlook for the future

On 31 December 2020, the DBT released the draft of a new National Science, Technology, and Innovation Policy (STIP) for public consultation (Department of Science and Technology, India, 2020). This is the fifth release of India's national science and technology policy and is expected to set the trajectory of Indian scientific development in the mid-to-long-term.

Similar to China, scientific planning in India used to be synchronous with the national socio-economic Five Year Plans. Such practice stopped after Prime Minister Narendra Modi assumed office and replaced the Planning Commission with a national policy think tank, NITI Aayog (short for National Institution for Transforming India). Nevertheless, each ministry was still expected to continue with making their individual five-year plans (Tiwari, 2019). In addition, the Indian government periodically launches its long-term vision of science through dedicated major planning initiatives.

A brief review of India's national science policies may help to put the 2020 STIP in context. The first national policy on science was adopted in 1958 and laid the foundations of Nehruvian science to 'cultivate scientific enterprise in pure science research, applied science research' and the promotion of 'scientific temper' through education and training programmes (Government of India, 1958). Whereas the 1958 policy focused on 'science', there was a clear turn to 'technology' in the second policy, the Technology Policy Statement, adopted in 1983. At the time, India had developed a good industrial and agricultural base and was ready to shift its attention to promote indigenous development so as to benefit the masses. This was also when India set out its global ambitions and identified information, electronics and biotechnology as strategic investment areas (Government of India, 1983). It was not until 20 years later that India set its third national policy on science. The Science and Technology Policy 2003 recognised that research and development (R&D) had become ever more interdisciplinary and required building of clusters of institutions and multinational collaboration. Pressured by its rising neighbour China, India pledged to increase its spending on R&D to 2% of its GDP, although this target has yet to be met at the time of writing. The fourth policy, Science, Technology, and Innovation Policy was released in 2013, when the Bt crops dispute was reaching its climax. It aimed to establish an 'aspiring India' where 'science technology and innovation for the people' was officially adopted as the new paradigm (Ministry of Science and Technology, India, 2013: 1–3). As discussed in the next chapter, the 2020 STIP reinforced the ethos of a 'people-centred' scientific vision and pledged expansion of science training, education and spending. One of the key new initiatives was to promote open science so as to further incentivise creativity from the bottom-up (Department of Science and Technology, India, 2020).

In fact the draft STIP opens with a full chapter outlining how the public will be guaranteed wide and free access to research. Domestically, it has been proposed that a National STI Observatory would serve as a publicly accessible repository to life science research. This is complimented by a new dedicated portal, the Indian Science and Technology Archive of Research, which provides access to outputs of all publicly funded research. In addition to promoting free circulation of knowledge domestically, internationally, the government has pledged to improve the global visibility of Indian journals so as to enhance the presence of Indian (institutional) research in the world (Department of Science and Technology, India 2020). The most radical proposition was the 'One Nation, One Subscription' initiative programme. The intent was for the government to negotiate with journal publishers for a nationwide subscription so that more than 1.3 billion people may have access to articles (Mallapaty, 2020).

The discussion of 'One Nation, One Subscription' emerged in the second half of 2020. It was a reaction to the global Plan S initiative, which is championed by European funding agencies. The Plan S mandates that works funded by these agencies will be made freely available to the public immediately upon publication (Van Noorden, 2020). While many European funders can cover article-processing charges imposed by journals, this may not work for India, a country already squeezing its pocket for R&D. It thus opted to negotiate its own national-level subscriptions with big publishers so as to provide free access to its citizens. If successful, this would be even more progressive than Germany's current arrangement in which nationwide subscriptions only apply to academic institutions.

Nature correspondence speculates that being the world's third-largest producer of science and engineering articles, India 'might carry more weight in negotiations with publishers' (Mallapaty, 2020). Some low- and middle-income countries may look to India as a role model to follow. However, open access advocates criticised such moves, arguing that India's move will only re-affirm a subscription-based culture rather than joining the global movement of completely moving away from it (Koley, Goveas and Chakraborty, 2020). One expert in Canada criticised such a move, stating it would only 'encourage everyone to fend for themselves' (Mallapaty, 2020).

It seems that, once again, India has signed up to the general idea of how science should be governed – the promotion of open and dynamic knowledge circulation – but once again chose to take a slightly different path. Both the criticism over India's short-sightedness in opting to compromise with national subscriptions, and the praise over its radical yet pragmatic proposal make sense. The difference lies in from whose perspective one is looking and what values one thinks this proposed subscription subscribes to. Does it underline more of a national interest or just a commitment to openness? Is such a proposal essentially inward-looking or future-oriented?

We are not here to provide a judgement on this. In fact, as the plan is yet to be finalised and delivered, it may be accepted or rejected by Indian society or by major publishers for a completely different set of reasons. Regardless of whether India joins the global initiative or not, global discourse and capacity on promoting open access will be affected by India. Similarly, India's narrative, and the 'weight' it carries, will not only be determined by the quantities of its publication, but also by the evolving opinions on what a productive path is. Thus, what we are certain of is that much of the debates both in and outside of India are taking place concurrently, and they will feed into and respond to each other.

Regardless of the outcome, in the latest STIP, there is a clear continuation and expansion of the government's support for bottom-up innovations. Open science is essential to empower the masses.

It perhaps will further expand the range of actors that 'could do science' or expand the range of research one could do outside their expected role or expertise. It highlights the imperative that a parallel upgrade of governance is also needed. One can no longer sit in the comforts of the 'bureaucratic amplification of credibility', but must reconsider how bureaucracy itself can maintain credibility and relevance to the changing landscape of science.

5

The dragon–elephant tango: making sense of the rise of China and India

There are two ways of interpreting the 'dragon–elephant tango' (*longxiang-gongwu*), both of which this chapter addresses. At one level, the 'tango' is an analogy of China and India's co-dependence and long-term cooperation, a political rhetoric echoed by successive leaders since the 1980s (Feng et al., 2014; Ma, D., 2019). But on another level, the phrase is also a good description of the 'friend-or-enemy' duplicity China and India projected to not just the West, but to Global South countries too.

For those with more positive views, the rise of two Asian powers raises hope for expanded resources to counter global health inequalities. India has been reverse-engineering patented drug molecules and exporting huge volumes of high-quality affordable medications worldwide since the 1970s. Despite its 2005 amendments to the patent law after joining the World Trade Organization (WTO), domestic market competition and government control over patent extensions and prices have enabled it to maintain its mantel of 'the pharmacy of the Third World' (Duggan, Garthwaite and Goyal, 2016). COVID seems to have heralded a new era of response to health crises, with the Global South increasingly stepping in to address resource inequalities. In early 2021, while the UK and the EU were in a row over vaccine export controls, one Australian diplomat concluded, after visiting vaccine manufacturing sites in India, that 'there's only one nation that has the manufacturing capacity to produce sufficient quantities to satisfy the demands of citizens in every country, and that's India' (Peel et al., 2021; Roy, Rocha and Das, 2020). In fact, as it contributes to 60% of the world's vaccine production, India is set to become a benchmark in vaccine distribution, using technology to ensure targeted and phased distribution (PTI, 2021). Similarly,

China has tried to establish itself as an alternative innovation hub for the developing world. For China, the global vaccine race started in 2006 when it competed against the US in developing a vaccine for Ebola, a virus that had no approved treatment and plagues West Africa (Carlson, Reiter and Lu, 2018; Gan, 2017). In 2017, China's State Food and Drug Administration approval of China's CanSino Ebola vaccine coincided with the Chinese Communist Party's 19th National Congress. 'Chinese' ingenuity was celebrated in helping to address a public health threat in Africa and highlighted as a testament to China's role as a responsible nation, committed to improving global 'common (health) security' (Gan, 2017). The ambition of using science both as a status booster and a tool for global diplomacy was echoed in the headline of CanSino's double-page advertisement in *Nature*, which describes CanSino's work as 'a best shot at global public health response' (CanSino *Nature* advertisement). The CanSino vaccine potentially offered a more practical solution in terms of logistics. This is because, similar to the COVID-19 vaccine, Chinese researchers used traditional inactivated vaccine paths in which the resulting vaccine could be stored at 4°C for a long time, and remained stable at 37°C for about three weeks, whereas the WHO-approved vaccine, developed by the pharmaceutical company Merck and funded by the Canadian and American government, had to be stored at –70°C or below, and was only stable for one week at 4°C. But to date, this Chinese Ebola vaccine has not yet received other international approval. As discussed later in the chapter, this has shaped China's strategy for its COVID vaccine development.

But there also seem to be legitimate reasons for caution. Scepticism towards China's and India's rise is not limited to the West. The two countries' leadership and trustworthiness are also contested by their regional neighbours (Flemes, 2009; Vieira and Alden, 2011). While it is true that emerging economies, such as the BRICS countries, are replacing some North–South collaborations with South–South collaborations given their promises of tackling shared development and scientific issues, there are also political rivalries amongst these emerging powers, which puts a question mark over their ability to ameliorate chronic global inequalities or to empower the Global South (Thorsteinsdottir et al., 2011). Taking vaccine uptake as an example, availability and accessibility to vaccines are not necessarily translated into actual inoculations (WHO, 2019). At the beginning of 2021, the intensity of China, India and Russia rushing to mark their influence with their respective vaccines may make one wonder if a regional vaccine map would be an instructive visual aid to a student of international relations seeking to grasp geopolitical fault lines and divisions. Emerging health diplomacy from the Global South may not necessarily improve either global health or international relations, but, as this chapter explains, it may create a more fragmented world and worsen geopolitical divides.

This chapter sheds new light on how to make sense of the sociopolitical complications and entanglements of China's and India's rise in the life sciences. While nationalism is often employed as a master frame to understand the development of science in China and India, we contest that a 'nationalist' lens alone provides only a partial, if not reductionist, view of the rise of the two countries. This is because it diverts our attention from how the two countries align national priorities with memories of colonialism, which has a powerful appeal to the Global South. A nation-state lens also tends to render a homogenised imagination of 'docile' peoples. It blinds us from seeing some of the key challenges China and India are still struggling with at the social level and the real limit of their outreach.

We hope to invite and enable you to 'think *from and with*' the two countries (Walsh, 2018) so as to understand the apparent contradictions embodied by them. We draw attention to an often ignored aspect of their shared political logic, namely how the shared imaginary of the 'developing world' has fed into a leftist science populism. It is important to note that leftist science populism does not negate nationalism, but it helps to contextualise how national ambitions are articulated. The concept of a *leftist science populism* is explained in the next section. In short, it can be explained in three steps. 1) It is populist in that the political legitimacy of a scientific agenda is founded on the basis of a 'people vs elites' dichotomy (Mudde, 2007). That is, a populist science governance builds on an epistemic power imbalance between 'the people' and (Western) scientific hegemonies. The expansion and development of scientific agendas, as such, is focused on converting scientific possibilities into visible benefits by the people, for the people. 2) The anti-hegemonic element in this developmental view of science makes it different from right-wing populism as it draws on a vertical distinction between the people and the elites, rather than a horizontal distinction based on national membership as most right-wing populisms do (De Cleen and Stavrakakis, 2017). 3) We also draw attention to the transnational nature of this leftist view. As this chapter demonstrates, both China and India do not have an exclusionary view of the 'people' in their scientific vision. Quite the contrary, both countries promote their scientific ambition with an inclusive view by drawing on 'the legacy of subalternity' (Halliday, 2016). That is, the development on science and technology, and more importantly, the right to steer such developments, are seen by Global South communities as being critical to lifting societies out of their socio-economic, and by extension, political subaltern status. Both China and India, in establishing and maintaining their role as representative leaders of developing countries, draw on this collective sentiment. However, through the examination of the 'dragon–elephant tango', or the entangled scientific development of China and India, this chapter demonstrates that an internal irony of leftist science populism is that it transforms subaltern initiatives into hegemonic ones. This helps to make sense of its conflicting reception in the Global South and why, as exhibited in the case of responding to COVID, it may be counterproductive.

In what follows, we first unpack the concept of leftist science populism. Similar to all policies, national science strategies embody the confluence of various political concerns and rationales, and cannot be reduced to only leftist populism. Yet, it helps to provide a more realistic view of how Global South societies manage the dual task of modernisation and globalisation and how they identify, align and justify national development trajectories on the basis of their position in the international power paradigm (Zhang, H., 2018). In the second section, we review China-India collaborations (or lack of) on science and science policy in their shared ambitions of becoming a global player. Leftist science populism helps to contextualise the gap between the two countries' official views and actual practices in R&D exchanges and the latent effect of the two countries' global expansion. Finally, we examine the latest critical event for both countries, that is (at the time of writing) the ongoing vaccine diplomacy during the COVID pandemic. Whereas the Chinese strategy can be summarised as contrast, collaborate and calumniate, the Indian approach is characterised as contest, convert and control.

We point out that the logic of leftist science populism exhibited in both China's and India's vaccine diplomacy further fragment, rather than unite the world. As the global public is segmented by the technical options that they were allowed to take, it erodes public confidence and trust in emerging technologies.

By unpacking some of the policy and social contradictions evoked by the 'dragon–elephant tango', we offer a new approach to evaluate the strength and limits of these rising nations. This leads to more productive ways to maximise the benefits of technical solutions while maintaining public trust and confidence in science, which will only play a greater role in international relations and global responses to public welfare.

Leftist science populism

Populism is a contested concept, for real-world populism can often accommodate a mix of contradictory political impulses (Aslanidis, 2018). Its various definitions are closely associated with the fact that there are many variants of populism, For example, it could be protectionist, redistributionist, neoliberal, anti-liberal, national or transnational (Brubaker, 2017; De Cleen et al., 2017; Ivaldi, Lanzone and Woods, 2017). Populism is not necessarily demagoguery either, for its intentions can be aimed at mass mobilisation or de-mobilisation, de-institutionalisation or re-institutionalisation of power structures (Barr, 2009; Brubaker, 2017; Mudde, 2016). In short, the 'actual contents' of populist politics may vary greatly, yet at its core, the term populism refers to a 'logic of articulation' in political rationales (Laclau, 2005b: 33).

The commonly agreed minimalist definition of populism is the juxtaposition between 'the people' and 'the elites' in legitimising political actions and steering public opinions (Ivaldi, Lanzone and Woods, 2017; Mudde, 2007). By speaking to the sentiment of the 'plebs', or the socio-politically disempowered, and advocating a redistribution and re-democratisation in collective affairs, populism does not limit itself to matters of socio-economic equality. It is also a way for subaltern groups to voice demands for respect and recognition withheld from them (Brubaker, 2017; Caiani and Padoan, 2020; Hochschild, 2016).

Based on this minimalist definition, there are a few further distinctions that need to be made to clarify what we mean by leftist science populism. First is the critical conceptual difference between populism and nationalism. Populism, especially right-wing populism as we've seen in the US under Donald Trump, often evokes nationalism. But they are not synonmous. For the purpose of our discussion, it is safe to say that populism has wider grassroots appeal than nationalism. It may coincide with or go beyond nationalist agendas, depending on how 'the people' is defined (De Cleen and Stavrakakis, 2017; Moffitt, 2017; Stavrakakis et al., 2017). As Argentinian political theorist Ernesto Laclau rightly argued, 'political practices do not *express* the nature of social agents, but instead, *constitute* the latter' (Laclau, 2005b: 33, original emphasis).

This point is perhaps better explained through a comparison between right-wing and left-wing populist stances. Right-wing populism often defines 'the people' along ethno-nationalistic lines, whereas left-wing populism identifies 'the people' by their socio-economic class or political status, such as their positionally in an epistemic hegemony (Caiani and Padoan, 2020). To put it another way, the difference between right-wing and left-wing populism lies in their respective horizontal and vertical construction of oppositions. Right-wing populist rhetoric is rooted in an 'In/Out' horizontal opposition between ethno-national identities, whereas left-wing populist draws on a vertical 'Down/Up' distinction of the positional difference in the hierarchy of power, recognition and/or socio-economic capacities (De Cleen and Stavrakakis, 2017). Thus, by 'leftist science populism' we refer to a populist vision of scientific development that speaks to the vertical power imbalance between the communities in the Global South and the Global North which have had overwhelming dominance over scientific resources.

The conceptual difference between populism and nationalism cannot be over-emphasised. Given right-wing populist dominance in Western discourse, it is easy to be blinded by a fixation on the nation-state. Consequently the role of populism in shaping transnational governance often gets lost in the discussion (Holliday, 2016; Plagemann and Destradi, 2019). For example, the term 'medical populism' was proposed in 2019 by Gideon Lasco and Nicole Curato. They introduced the concept as 'a political style that constructs antagonistic relations between "the people" whose lives have been put at risk by "the establishment"' (Lasco and Curato, 2019: 1). While the authors noted that 'the establishment' could 'range from the state to medical experts, to big pharma to "the West"', their choice of examples are predominantly from right-wing populism, such as the 2003 polio vaccine boycott in Nigeria, or 2016 anti-vaxxer MMR movement in Italy. Their analysis focused on the political strategies of dramatising a crisis, a 'horizontal' othering of nation states, and forging knowledge claims so as to protect 'the people' from 'outsiders' (Lasco, 2020; Lasco and Curato, 2019: 3; Lasco and Larson, 2020). But not all populism builds its authority over conspiracy theories and fake news. This leads to our second point.

A second difference that can be drawn between right-wing and left-wing populism is the source of legitimacy and justification of entitlement. As much discussion on populism to date has been Western-centric, there is a tendency to reduce populism to its 'exclusionary right-wing form' (Caiani and Padoan, 2020; Holliday, 2016; Plagemann and Destradi, 2019). The resurgence of populism in many Western developed countries is often associated with the negative consequences of economic globalisation (Rodrik, 2018). It is not surprising, then, that such populism often has a chauvinist character and explicitly or implicitly provokes racial or socio-cultural superiority. But there is also an inclusionary left-wing variant of populism in which political ambition is projected as striving for a more level playing field for all. This is the struggle for recognition of the needs and desires of all 'the people', who may live beyond particular national boundaries but have equal entitlement to recognition (De Cleen et al., 2020: 149). This cosmopolitan vision of the populist base enables leftist populism to simultaneously relate to domestic and international audiences. One example of such transnational leftist populist project is the Democracy in Europe Movement 2025 (DiEM25) (De Cleen et al., 2020). Initiated in 2016 by a group of luminaries such as philosophers, Slavoj Žižek, Srećko Horvat and former Greek Finance Minister Yanis Varoufakis, DiEM25 constructed 'the people' as ordinary citizens across the European continent and pitted them against 'unaccountable "technocrats"' who mask 'highly political, top-down' decision making processes as '"apolitical", "technical", "procedural" and "neutral"' (DiEM25, 2016: 2). It pledges to expose and end 'relations of power masquerading as merely technical decisions' (DiEM25, 2016: 7–8).

Building on the two points made above, the third point relating to leftist science populism is its potential transnational impact (De Cleen et al., 2017). A simple cosmopolitan vision of 'the people' itself does not make a political strategy populist, nor does it make populist rhetoric potent. Rather it is the fact that the potential audience are united through their imagination of what Laclau (2005a: 73) termed 'an equivalential relation' between their respective demands. It is 'an equivalential articulation of demands making the emergence of the "people" possible' (Laclau, 2005a: 74; Aslanidis, 2018). For many Asian communities, populist ideals are rooted in their colonial memory. The appeal of populism is a 'legacy of subalternity', a yearning to counter-balance the epistemic and political dominance of the West (Holliday, 2016). As the next section demonstrates, there is a clear counter-hegemonic tone that runs through both Indian and Chinese scientific strategies. While there is a strong recognition of domestic needs, national visions of research development in these two countries were hardly ever restricted to the wellbeing of 'Indian' or 'Chinese' people. Rather they are representative of the globally disadvantaged in general, and there has always been an effort to make transnational appeals to developing countries' experience in general. In other words, the building of China's and India's research capacity in the contemporary world is also a constitutive process of the 'people' they and their science aim to speak for.

There is also a subtle yet noteworthy difference between '*trans*-national' populism in which the political construct of the 'people-as-underdog' supersedes the boundaries of the nation-state and '*inter*-national' populism which calls upon the coalition of nationally bounded people-as-underdogs (De Cleen et al., 2020). The latter instrumentalises through 'a populist chain of equivalence', which brings different groups together, and unites their different pursuits through a shared opposition to global or foreign elites in general (Moffitt, 2017). As the next section demonstrates, the trajectory of China's and India's scientific developments are enmeshed with both transnational and international populist logic. They are entangled and easily slip into one another. But it is precisely this duality that Chinese and Indian instrumentalists use to appeal to both their constituents (the sovereign people) and a wider global audience (the people 'at large').

Table 5.1 Key differences between right-wing and left-wing populisms

	Right-wing populism	Left-wing populism
Premise of 'the people'	Nationally or ethnically bonded	Not limited by nation-state boundaries
Framing of political opposition	'In/Out', horizontal	'Down/Up', vertical
Source of legitimacy	Dramatisation of outsider threat	Seeking fairness and equality
Justification of entitlement	Exclusionary, chauvinistic	Inclusionary, struggle for recognition
Intended outcome and latent effects	Mainly national	Could be national and/or transnational

In short, by leftist science populism, we refer to a political outlook that pits the people in the Global South against Global North authorities who undermine their freedom and/or ability to explore, apply and promote scientific possibilities that are seen as central to a good and respected life. The difference between left-wing and right-wing populism is summarised in Table 5.1. It is also important to be reminded that populism does not mean 'popular', nor does it automatically receive indiscriminate public support. Indeed, a successful delivery of populism requires charismatic leadership, which is not necessarily limited to the charisma of an individual, but relies more generally on soft power. Furthermore, one irony of populism is that while it starts as speaking for the subaltern group, it is often directed at replacing an existing hegemon with a new one (Holliday, 2016). Populist leaderships are often trapped in the logic that 'they, and they alone, represent the people' (Müller, 2016: 3). These limitations are exhibited in both China's and India's construction of their global leadership in science and in their vaccine diplomacy.

The choreography of China–India science relations

There is simultaneously a lot and very little that can be said about China–India science relations. There is a lot to be said because the two countries share similar subaltern aspirations of achieving national revival through science and technology. Leo Tolstoy famously began his novel *Anna Karenina* with the statement, 'All happy families are alike; each unhappy family is unhappy in its own way.' Yet for China and India, at least in the realm of science and technology, it is tempting to argue that the reverse is true. For while 'happy' or developed science in Western countries, such as the US, the UK and Germany, may have their distinct strengths and culture, over the larger part of the past 80 years, the 'unhappy' or underdeveloped scientific programmes in China and India have been confronted with similar sets of challenges: lack of funding or funding sources, global brain drain, bureaucracy, cultures of nepotism and rule by seniors, poor enforcement and the image of being an imitator rather than innovator of science (see Cao and Suttmeier, 2001; Desiraju, 2012; Mani, 2013; Padma, 2015; Prasad, 2005; Sabharwal and Varma, 2017; Zhang, 2011). In the 1980s, just as both countries were starting to revamp their scientific agendas, Chinese leader Deng Xiaoping reaffirmed with his Indian counterpart Rajiv Gandhi that 'the Asian century will only come when both China and India have developed' (Deng, 1993: 282). Deng pointed out further that both countries' developmental vision should not be constrainted to a nation-state, but needed to be 'elevated to the level of all of humanity' (Deng, 1993: 283).

China-India scientific collaborations as a bilateral national agreement can be traced back to 1988. A Joint Committee on Scientific and Technological Cooperation was set up by the two countries to plan, coordinate, monitor and facilitate cooperation in strategic areas. Biotechnology was identified as one of the eight 'priority areas' for co-development during the third meeting of the Joint Committee in 1993. More recently, as a counter narrative to the North American strategy of 'China plus one', which aims to divert Western manufacturing reliance (including the manufacturing of medical supplies and generic drugs) from China to other developing countries, Xi Jinping has proposed an expansion of the 'China-India Plus' cooperation on both strategic foreign policy issues and on practical cooperation 'in new energy, science and technology innovation' and the establishment of India-China High Level Mechanism on Cultural and People-to-People Exchanges (Prasad, 2018).

Yet, there also seems to be very little that the 'dragon–elephant' alliance has added, either to their respective domestic or to the international discourse on life science governance. Despite official recognition of shared developmental aspirations and needs, in reality, mutual scepticism towards the other as a new, rival hegemon has led them to dance around some of the shared regulatory and capacity building issues that have been plaguing their international image. Much of the scientific exchanges are arguably entrepreneurial, with pragmatic economic returns. For example, 2020 marked the 70th anniversary of diplomatic relations between China and India. Of the scientific collaboration highlights listed on the Chinese Embassy in India website were: two China-India Science, Technology and Innovation Collaborative Research Forums; three India-China Technology Transfer, Collaborative Innovation Investment Conferences; three India-China entrepreneurial information corridors; and six ministerial-level strategy setting conferences on science (Embassy of PR China, 2020). Yet the actual substance and the level of exchange of these 'highlight' events are disappointing. Both of the China-India Science, Technology and Innovation Collaborative Research Forums took place in Beijing. At the first forum in 2016 only two Indian nationals attended. Although one of the two attendees was the Director of National Institute of Science, Technology and Development Studies (NISTADS), Prashant Goswami, the forum was an individual-driven project rather than a formal national collaboration (Zhu, 2016). The second forum, which took place three years later, had 9 Indian attendees, led by another NISTADS director (CASTED, 2019). In comparison, the three India-China Technology Transfer, Collaborative Innovation Investment Conferences had attracted wider participations, each with around 150 attendees. This was a civil initiative focused on the marketisation of technologies and founded in 2016 by the India China Trade Centre in New Delhi and Yunnan Academy of Scientific and Technical Information in China (China-South Asia Technology Transfer Center, 2020). Despite being a business-oriented meeting, representatives from the Ministry of Science and Technology and the Chinese Embassy in India also attended these events (China-South Asia Technology Transfer Center, 2020). Given that the last of the 6 ministerial-level meetings on science policy took place in 2013, it seems that, since then, both Chinese and Indian policy makers seem to be more interested in commercial exchanges rather than engaging with governing issues.

Despite the absence of a strong bilateral engagement of science, a 'choreography' can be drawn, as China and India have both served as a mirror to each other. There is a noticeable leftist populism underlying both China's and India's latest national scientific strategies, namely China's Fourteenth Five Year Plan and India's 2020 draft plan of the Science, Technology, Innovation Policy (STIP) (Department of Science and Technology, India, 2020; Xinhua, 2021). The overarching tone and priority in both countries' agenda is to prioritise scientific and technological 'self reliance' so as to counterbalance international hegemonic powers. For China, the pressure comes mostly from the US, while for India, China's rise as a regional hegemon is arguably a more imminent worry. Moreover, both countries resort to achieving self reliance through empowering mass innovation and through 'game-changing' plans to expand respective 'global academic and entrepreneurial network(s)' (Mehta and Gopalakrishnan, 2021; Xinhua, 2021). India's latest national STIP was considered as 'In a true sense … a STIP by the people, for the people and of the people at large' (Mehta and Gopalakrishnan, 2021). It aims to continue its reputation as the land of grassroots innovation and enables this by having an all-encompassing Open Science Framework accompanied by a government negotiated 'one nation, one subscription' deal with journal publishers to set up free access to research across the world to Indian citizens (Department of Science and Technology, India, 2020: 13). Similarly, China boosts its policy support for grassroots funding and participation of translational research. Echoing India's engagement with its diaspora, China also aims to reverse its brain drain by setting up favourable and supportive measures to help foreign experts settle back in China. It further supports science and education in developing countries and commits to 'design[ing] and initiating major international projects and programmes on big science' and to further opening up national funding schemes to global talents (Xinhua, 2021:15–18, 103). Importantly, both India and China see their scientific rise as a mission, an equalising force in an uneven world. Both believe that their mission necessarily ties into and benefits not just constituencies at home, but empathic audiences across borders.

In addition, China and India observe and react to one another. For most of the twentieth century, India had the lead in science and science governance over China. While India had its first Nobel Prize-winning scientist in 1930, it took another 85 years for China to have its own Nobel Laureate scientist. Political rivalry aside, in the eyes of Chinese scholars, the academic legacy from India's colonial era, along with its wide use of the English language, provided Indian science with a significant advantage in internationalisation. When China turned its attention to facilitating industrial translation of applied research, Chinese scholars were envious of India's early start in embedding marketisation in its science strategy in the 1960s (Wang and Feng, 2018). By the 1990s, India had already established international collaborative networks in the pharmaceutical sector. Based on statistics released by the Union of International Association, Chinese scholars pointed out the fact that India is 'the most active developing country in joining international forums', which includes both inter-governmental organisations and global civil societies (Zhang, Q., 2020: 30). Chinese academia also tend to see Indian's research policies as more socially considerate. For example, India's 'inclusive innovation' directive which stresses gender balance and inclusion of the socially disadvantaged in national science and technology strategy has been highlighted as a model for sustainable development (Hu, Guo and Feng, 2016; Zhao and Chen, 2015). While China has pockets deep enough to attract overseas returnees, some argue it lacks the multi-layered civil initiatives in attracting overseas returnees such as India's previous short-term 'Know India Program' and Study India Programme, which help diaspora youth (e.g. children of overseas return scientists) to adjust to life in India (Hu, Guo and Feng, 2016). Before China kicked off its 'mass entrepreneurship and millions to innovate programme' in mid-2015, ideas such as 'Jugaad innovation', 'grassroots to global' and 'frugal engineering' were arguably Indian imports (Feng et al., 2014: 190; Hu, Guo and Feng, 2016; State Council, China, 2015; Zhang, Q., 2020).

India was also a leader in forming South–South collaboration. The founding of the India-Brazil-South Africa (IBSA) Dialogue Forum in 2003 was one of the most notable rising global power blocs during the first decade of the twenty-first century (Stuenkel, 2014). Science and technology is one of its priority areas. When evaluating IBSA's work on science and technology, the New Delhi-based think tank Research and Information System for Developing Countries

(RIS) first quoted political scientist Andrew Chadwick to highlight the role of science and power hierarchies: 'Globally, technology structures societies and global interactions by creating hierarchies of power between the haves and have-nots, suppliers and users, and between states and market-driven multinational corporations' (Chadwick, 2006). The report went on to stress that 'for developing countries the technological dynamism per se is meaningless unless ... it enables them to translate the gains from S&T into economic gains' (RIS, 2016: 33). This is seen as a tactic to gain popular support of IBSA's lead in global innovation and to compete with, if not to subvert, the dominance of the 'mega science project' driven innovation paradigm dominated by the US and EU (RIS, 2016: 37).

However, a question mark hangs over the actual impact of the IBSA initiative. For one thing, there was a lack of funds. In 2005, the coalition launched the Rio Declaration of Science and Technology and subsequently launched joint funding to support trilateral cooperation. However, as the fund was only $3 million in total, it was understood more as a symbolic gesture (Stuenkel, 2014: 76–7). This also constrained the formation of new research cultures within the group. For example, the number of intra-BRICS academic exchanges is far lower than between IBSA and developed countries. According to the RIS report, as of 2016, only 300 researchers had benefited from the coalition (RIS, 2016: 33). To some extent, the IBSA collaboration resembles more of an *inter*-national initiative rather than a *trans*-national one. The IBSA Working Group of Health admitted that the coalition's value lies more in spirit rather than making actual change, for members 'do not have adequate time and incentive truly to understand other member countries' approaches' (Stuenkel, 2014: 74). An additional challenge is that, as seen from the West, IBSA resembles a revolutionary force from the Global South; and, in the eyes of their regional neighbours, such as China, Pakistan, Argentina and Bolivia, their leadership and trustworthiness is far less than clear (Flemes, 2009; Mokoena, 2007; Vieira and Alden, 2011).

A turning point in the India-China scientific race came in 2012, when the then Prime Minister, Manmohan Singh, admitted to the delegates at the 99th Indian Science Congress in Bhubaneswar that 'over the past few decades, India's relative position in the world of science had been declining and we have been overtaken by countries like China' (Jayaraman, 2012). A key reason for India's decline in relative competitiveness was a lack of financial and human resources. China's gross domestic expenditure on research and development had enjoyed approximately 20% annual growth since 1999 and become the world's second largest research and development (R&D) expenditure in 2009. In contrast, despite repeated government pledges to increase India's R&D expenditure to 2% of its GDP, the actual figure lingered around 0.6–0.8% (Mehta and Gopalakrishnan, 2021). In comparison, Russia and Brazil achieved an R&D budget between 1.1 and 1.25% of their GDP at the time (Joseph and Robinson, 2014). A deficiency in infrastructure investment in science also effects education and job opportunities. India has one of the lowest density of scientists and engineers in the world (Padma, 2015). Thus, while in 2020, Chinese research institutions were tasked to limit, if not shrink, their full-time staff, India's new national STIP aims to double the number of full-time equivalent researchers (Department of Science and Technology, India, 2020; Ministry of Human Resources and Social Security, China, 2019).

Singh's 2012 admission fuelled the ambition of Indian officials to adopt a 'warlike' attitude in making India a global scientific leader by 2030 (Jayaraman, 2012). Yet, also in 2013, China launched its Belt and Road Initiative (BRI). Initially proposed as an economic development initiative along landlocked Central Asian countries, the initiative has arguably became the most notable South–South collaboration in the second decade of the twenty-first century (Xinhua, 2013). In short, 'Belt' refers to China's ambition of expanding economic co-development westward with countries along the historical Silk Road, whereas the term 'Road' builds on the fifteenth-century maritime expedition routes of Admiral Zheng He, and refers to the Indo-Pacific sea routes that connect Southeast Asia, the Indian subcontinent, Western Asia, the Middle East and Africa (Yan, 2017). China has signed 46 agreements of cooperation in science and technology with BRI countries, and launched the China-ASEAN and China-South Asia science and technology partnership programmes (Office of the Leading Group for Promoting the BRI, 2019). India has always been sceptical, if not hostile about China's expansion of its geopolitical dominance in South Asia. In effect, BRI encircles India with Chinese influence. India thus boycotted the Belt and Road Forums in 2017 and 2019. Yet some scholars in India have urged the government to shift from an 'overly securitised posture' to China's rise towards 'an order-building approach' (Saran, 2019; Singh, 2019). Part of the reason is China's expansion on science and technology and the worry that India may miss out, both in terms of participating in joint research and in taking advantage of human capital (Masood, 2019).

A key difference between the BRI and the IBSA initiative launched a decade earlier is that China has been more aggressive in developing a transnational community with at least shared technical norms. In contrast to India's focus on delivering distance education programmes (Trines, 2019), China has been more attentive to integrating researchers and supporting staff along the BRI into China's scientific culture. For example, the Chinese Academy of Sciences has invested in 9 research and development facilities in BRI partner countries and has undertaken the training of 5,000 students in science and technology courses (CAS, 2019). In 2017, 38,700 students from BRI-associated countries studied in China through scholarships provided by the Silk Road programme, accounting for 66% of all students receiving such scholarships (Office of the Leading Group for Promoting the BRI, 2019). In 2018, 500 young scientists were offered short-term fellowships which enabled them to carry out research in China. In addition, more than 1,200 science and management professionals from BRI countries were trained in China (Office of the Leading Group for Promoting the BRI, 2019). The person-focused nature of China's South–South scientific community building holds immense promise because the primary driver of most research collaboration are the scientists themselves. These exchange programmes help to nurture a transnational scientific culture. In 2018, Joy conducted a focus group study of African students in Changsha. The participants were undertaking science and engineering degrees with Chinese scholarships. When asked if they would have preferred to study in Western institutions, such as in the UK or France, if provided with the same level of sponsorship, the African students unanimously said no. One of the participants studying earthworks said, 'Why would I? China offers the most practical solutions for me to use back home. Plus, we developing countries have a similar approach to solving problems.'

Another example is the Belt and Road Life Sciences Economy Alliance (www.brlsea.org), which has sprung into wider commercial and civil collaborations. The founding institution was not the Chinese government but BGI. Current members are a mixture of pharmaceutical companies, venture capitalists, research institutes, government and semi-government agencies from BRI countries (see www.brlsea.org/member). In the Alliance's first general assembly held in December 2020, a Belt and Road Regional International Standardization Committee was set up, announcing that 'a new round of globalization' had begun (LSEA, 2020). For 'the core of globalization is to create value and jobs in countries and regions', thus it is time to 'redefine retail, technology, production, and quality, including redefining standards, which will trigger an unprecedented revolution' (LSEA, 2020). In other words, the new round of South–South collaboration led by Chinese authorities and scientific communities is much more overt in its challenge of the global scientific epistemic status quo. It is more left-leaning and aggressive than IBSA and other South–South coalitions on science, for it is not simply framed as 'people in developing countries are underdogs' who need to catch up with the West, but challenges the norms of how standards are derived, and who has a say in creating them.

Of course, we want to be cautious not to over exaggerate the influence of China, for so far its (attempted) influence over transnational norms remains in the technical realm. This does not necessarily mean that China or Chinese science has acquired soft power or authority within the Global South. As the next section demonstrates, in fact China still has a long way to go to address its soft power deficiency. However, it would be equally unwise to overlook the evolution of India's and China's global vision or to fail to consider some of the latent effects of these initiatives. From their 1988 co-development agenda to their 2020 national scientific strategy, from IBSA to BRI, there is continuous if not progressive leftist vision in China and India. Both countries see their rise in global science as a mission, not simply for their own national constituencies, but also in trail blazing a more level playing field people in the Global South.

COVID vaccine diplomacy: the hard and soft power tango

The COVID pandemic is seen by many as a watershed event, which will revolutionise how we run the world and reset the global order

(Dunford and Qi, 2020). It may take at least a couple of decades for the full consequences of COVID to unravel. Among the many technical, economic and socio-political lessons that can be drawn from the pandemic, one of the major shocks to many communities is the virus's exposure of widening social inequalities within nation-states and across the world, as well as a consequential divergence in worldview. While people celebrated a succession of unprecedented vaccine developments under lockdown, there was also a sobering reckoning that, depending upon where one lived, the level of access and amount of choice available to these scientific possibilities varied.

Despite repeated appeals for global solidarity and a shared recognition that the world would not be safe while any single country was still fighting the virus, the race for inoculation further accentuated the socio-economic and epistemic divide between the Global North and Global South. Western countries were criticised for creating 'unprecedented inequality' by hoarding billions of doses of excess vaccines with an ambiguous timeline to share with developing countries (Al Jazeera, 2021). The US, for example, secured close to 3 billion doses of the COVID vaccine, which was eight times the total population, while Canada secured five doses per Canadian (*The Economist*, 2021; Rawal and Bai, 2021). Meanwhile the WHO created the COVAX programme (the COVID-19 Vaccines Global Access programme), it can only cover 20% of the population in most countries reliant on the programme (Chauvin, Faiola and Dou, 2021). Eighty-four of the world's poorest countries will need to 'wait for their turn' until 2024 to achieve mass COVID-19 immunisation (Safi, 2021). Not only are countries with shallower pockets and less diplomatic might squeezed out of the vaccine supplies, but this also re-affirms a more challenging issue: in the words of Chinese media, the world seems to be operating 'under two different skies' and thus unable to empathise with each other's priorities (Li and Guan, 2021). South Africa was asked to buy doses of AstraZeneca's vaccine at a price nearly 2.5 times higher than most European countries (Rawal and Bai, 2021). But when India and South Africa approached the WTO to waive parts of the Agreement on Trade-Related Aspects of Intellectual Property Rights (TRIPS Agreement), it was 'vehemently opposed' by wealthy nations like the US and Britain as well as the European Union' (DW News, 2021). While India and South Africa cite the reasons of 'timely access to affordable medical products' and the importance of 'scaling-up of research, development, manufacturing and supply of medical products essential to combat COVID-19', developed countries argue that this is 'robbing' pharmaceutical companies 'the incentive to make huge investments in research and development' (DW News, 2021). This difference between the privileged and the underprivileged was further illustrated by the leak that British Prime Minister Boris Johnson attributed vaccine success to 'capitalism' and to the 'greed' of large drug companies at a meeting with his parliamentary backbenchers (BBC, 2021).

A March 2021 World Economic Forum article argued that India and China were the solution to this North–South divide and that the two countries could end 'vaccine nationalism' altogether (Rawal and Bai, 2021). The article pointed out that vaccine nationalism was a phenomena of scarcity. Thus, ending global rivalry would lead to increased vaccine productivity. While the US was estimated to produce almost half of the world's COVID vaccines in 2021, Indian and China combined were projected to make up most of the other half. If the projected 7 billion does produced by India and China were distributed to the world, then in theory, almost everyone on Earth would receive at least one dose. While China and India have signed up to the WHO's COVAX programme, the authors of the paper were still puzzled: why has neither China nor India walked the talk and sold the majority of their capacity to COVAX at a fair price? (Rawal and Bai, 2021). The article concluded with an urgent reminder to both countries that they were facing 'a once in a generation opportunity to act as the leaders they aspire to be' (Rawal and Bai, 2021).

But even before India had to battle with a sudden domestic jump in coronavirus infections cased by the Delta variant in early 2021, neither China nor India had offered more of their vaccine production to the WHO. Why? The answer lies in the fact that the two World Economic Forum authors, both founders of North American health ventures, had the wrong idea about the kind of 'leaders' that China and India aspire to be. China and India no longer simply aspire to lead the 'world's factory', either for electronic gadgets or for vaccines. They have long shifted their aspiration to be the leader of world *innovation*. Both China and India saw the vaccine rollout as a 'once in a generation opportunity', not in selling their production capacity *per se*, but in promoting their scientific ingenuity and their own vaccines, which had yet to be approved by the WHO.

In fact, despite a global need for vaccines, both China and India struggled to acquire emergency authorisation, let alone full authorisation from developed countries in the spring of 2021. Despite the fact that India has long enjoyed a global reputation of being a quality vaccine *manufacturer*, its indigenous COVAXIN, developed by Bharat Biotech in collaboration with the ICMR, was considered as an inferior 'backup' and had a hard time winning recognition from the Global North. Frustrated by such charaterisations, the Chairman of Bharat Biotech lashed out at media: 'Why is nobody questioning the UK trials? Because Indian trials are easy to be bashed' (Pandey and Sharma, 2021). Similar to Chinese firms, Bharat Biotech also needed to build its vaccine's global profile by relying on support from the developing world. Developing countries like Zimbabwe, Iran, Nepal and Mauritius were the first to give COVAXIN the green light (Mitra, 2021).

Thus, for both China and India, their vaccine diplomacy is not solely driven by nationalism, but is also heavily shaped by a leftist populist logic, which enables them to project themselves as an alternative source of global leadership to rival the Global North's dominance. But, as demonstrated below, such populism is also counterproductive, for it further fragments rather than unites the world, and diverts the two countries from attending their own (soft) power deficits.

In what follows, we first examine China's and India's vaccine diplomacy. A leftist populist lens helps us to see beyond a simple national protectionist logic in the two countries' vaccine race, and sheds light on some of the more fundamental concerns these emerging countries have over scientific development in the global age. We then further interrogate the phrase 'the people' and analyse some of the shortfalls in China's and India's approach. Through the curious phenomenon of 'selective vaccine hesitancy', we underline our earlier point that an intrinsic irony of populism is that in striving for recognition it often deviates from recognising public needs. This is important not only to future global vaccination programmes, but also in how to improve technology uptake and public trust in science in general.

China's vaccine diplomacy can be summarised as contrast, collaborate and calumniate. The contrast between First World privileges and Third World worries forms a critical part of China's global outreach. Despite varied efficacy results of the Sinovac vaccine reported from Turkey, Indonesia and Brazil, Chinese media was adamant that Chinese vaccine 'performs better' than Pfizer's, for Pfizer's had 'a relatively steep price' and their storage requirements at ultra cold temperatures effectively rendered them a Western luxury whereas China's vaccines offered a more practical solution for poorer countries (Reuters, 2021; Zhang, H., 2021). Headlines which contrasted rich countries grabbing 70% of global vaccine supplies with China providing vaccine aid to 53 developing countries further galvanised China's image as champion of the Global South (Li and Guan, 2021; Wouters et al., 2021). More importantly, having learned from its failure in getting recognition for its Ebola vaccine research, China recognised the importance of international collaboration at an early stage (Jian, 2020). In fact, one of the five leading COVID vaccine teams was led by the military scientist Dr Chen Wei and CanSino Biologics, the same partnership that initiated China's Ebola vaccine project a decade earlier. In retrospect, the team considered that a key reason for their failure in acquiring WHO approval was their unilateral approach to clinical trials, whereas Merck's vaccine benefited from clinical collaboration and evaluation from different countries from early on (Jian, 2020). Thus, in addition to the fact that due to its effective control of the COVID virus, China would soon run out of patients for vaccine clinical trials, the inclusion of other developing countries in delivering clinical trials as well as vaccines was also of strategic consideration in beefing up the perceived validity and legitimacy of China's scientific advancement. China's other vaccine giants, such as Sinovac Biotech had streamlined its overseas clinical trials, local staff training and technology transfer to countries such as Indonesia, Malaysia, Brazil, Russia, and the UAE (China Business Weekly, 2020; Lei, 2021). As commented by the *Washington Post*, vaccine diplomacy could be a 'double win' for China: it opens new markets for its pharmaceutical products while building goodwill and expanding its influence transnationally (Chauvin, Faiola and Dou, 2021). But the most counterproductive element in China's vaccine diplomacy was its resorting to calumniation. China launched an aggressive disinformation campaign to undermine public trust in the vaccine developed by Germany's BioNTech with US pharmaceutical giant Pfizer. In a concerted effort, leading Chinese national papers criticised Western media's downplaying of possible serious side effects and urged Australia to halt its use of the Pfizer vaccine (Hui, 2021). This not only fed into the global anti-vax movement; the over-politicised language did not help with improving China's own scientific credibility. When China struggled to make an impact in South Asian countries, with Nepal, Bangladesh and Sri Lanka asking India for supplies of the British AstraZeneca vaccine instead, China blamed India for 'smearing' China's cooperation with these countries (Krishnan, 2021; Pasricha, 2021). China may not see itself as a regional hegemon, but its diplomacy seems to be trapped in the populist logic that 'they, and they alone, represent the people' (Müller, 2016: 3).

India's vaccine diplomacy has had a better global reception, partly because it has been much less overtly aggressive than China's. Its approach can also be summarised in three words: contest, convert and control. In comparison to China, India did not resort to media sensations and smearing campaigns in juxtaposition of 'the people's' rights to affordable scientific solutions and Global North's dominance. Rather, it contested the global status quo through the joint filing of the aforementioned WTO TRIPS waiver with South Africa in late 2020. It has since received backing from more than 80 WTO members. Although the proposal was rejected in January 2021, negotiation is still ongoing at the time of writing. Médecins Sans Frontières (2021) subsequently published a world map showing countries' positions on waiving monopolies for COVID-19 medical tools. This map is more powerful than any words in showing the North–South divide. The WHO chief Tedros Adhanom Ghebreyesus condemned wealthy nations' 'serious resistance' to this waiver (AFP, 2021). But for India, apart from former colonial powers, the shadow of a neighbouring hegemon, namely China, arguably generates more worry. Thus, India's 'Neighbourly Vaccine Diplomacy' or 'Vaccine Maitri' was pitched to curb China's influence and to convert some of the countries needing vaccines to its favour (Pasricha, 2021). India also aimed to convert its image from a scientific imitator to an innovator. Thus, with vaccine development, India had to keep a fine balance between being a major producer of Western vaccines and pushing forward its homegrown alternative. There was a telling inconsistency in India's emergency approval. At the end of February 2021, India approved the emergency use of its homegrown vaccine

Covaxin in the absence of clinical trial III data, and Covishield, a version of the AstraZeneca vaccine produced by the Serum Institute with incomplete immunogenicity data. Yet Pfizer, which would have passed the same yardstick, was not granted clearance (Kapur, 2021). But in early March Pfizer, who originally did not plan to manufacture its vaccine in India due to its exportation and pricing control, expressed to the Indian government its willingness to hand over some of the production to India, 'if faster clearance [and] export freedom' were assured (Arora and Das, 2021). The discriminating treatment between AstraZeneca and Pfizer arguably helped to shed light on the question posed by the World Economic Forum article we cited earlier. For India, the delivery of its vaccine diplomacy does not simply mean selling its muscles to the WHO, but an aspiration to have at least partial 'control' over the global landscape of what gets produced and delivered. While, arguably, India made an effective political decision in leveraging its global position, this did not help public confidence or trust. For a country that has never registered significant hesitancy toward vaccines, India's initial vaccine rollout was met with public scepticism, with 40% of its medical and frontline workers reportedly refusing injections (Pradhan and Sen, 2021). One key reason was a general worry that political discretion trumped scientific data, and a lack of free and fair public discussion (Pradhan and Sen, 2021).

Both China and India pitted the health and safety of the general public against the established global or regional hegemonies. Yet by doing so, they implicitly based their health diplomacy on a divisive vision of the world. Broadened technical availability (such as the availability of vaccines from different countries) may not necessarily translate into wider courses of action for they may be pre-emptively curtailed by smearing campaigns or deliberate regulatory blockage. More importantly, counter-power itself does not necessarily constitute power or even public credibility. While both China and India may have accumulated significant hard power in the discovery and production of science, in their strive towards global prominence, they are far from adroit in weilding soft power to gain confidence from and communicate to diverse publics.

The fact that both China and India struggled to have their indigenous vaccines recognised by developed countries is itself arguably a good testament to their credibility deficit. But their soft power also often falls short in connecting with the people in the Global South. We noted in an earlier section Bharat Biotech's frustration over the 'backup' characterisation of its vaccine; similarly Chinese vaccines were met with scepticism across the developing world. To be sure, vaccine hesitancy is a global problem and, to varying degrees, the credibility and trustworthiness of medical authorities poses a challenge to every society. But a lack of reputational and cultural capital to appeal to and reassure the public seem to be particularly damaging to China. Parallel to its vaccine diplomacy is a rise in what we call 'selective vaccine hesitancy'. That is, among 'the people' (mostly in the Global South) who are receiving China's vaccine, their resistance to vaccination is not due to a general scientific scepticism but is a reactive and selective rejection to a particular set of conditions in which the vaccine was prescribed to them (Wiyeh et al., 2018).

A 2020 survey showed that COVID vaccine acceptance was highest in Vietnam (98%), India (91%) and China (91%) (Wouters et al., 2021) The finding echoed a Wellcome Trust 2018 survey, which showed only 59% of Western Europeans and 72% of North Americans considered vaccines to be safe. In South Asia and East Africa, however, the respective figures were as high as 95% and 92%. Yet, in contrast to this pro-vaccine sentiment, was scepticism towards China's intention. According to a 10-nation survey conducted by the ASEAN Studies Centre at Singapore's ISEAS-Yusof Ishak Institute at the end of 2020, in the Philippines, Vietnam, Thailand, Indonesia and Myanmar, 63% of those surveyed had 'little confidence' or 'no confidence' that Beijing would 'do the right thing' for the global community (Seah et al., 2021).

In Pakistan for example, given the two countries' long-standing relationship, it was not surprising that the government was the first in the world to approve not only one, but three of China's COVID vaccines. Yet government endorsement did not translate into public trust. In the first two months of its vaccine rollout, when China's Sinopharm vaccine was effectively the only choice, uptake was low. For those who were 'offered the Chinese vaccine felt they are being given an inferior product' (Marlow, Mangi and Lindberg, 2020). Another example is the Philippines. An earlier survey had found that 94% of hospital staff were willing to take COVID jabs (Robels, 2021), but in March 2021, when the government dictated that hospital staff would only be given the Sinovac vaccine, the Philippine General Hospital's Physicians Association announced that 95% of hospital staff disapproved of receiving the Sinovac vaccine. This led the director of Philippine General Hospital to appeal to the public to 'separate the vaccine from our politics' (Robels, 2021).

Given how developed countries snapped up the majority of vaccines, China and India remain effectively the lifeline for most developing countries. Yet an often ignored fact is that resource limitations do not drain the public of their expectation nor capacity to demand accountable health care (Dube and Magalhaes, 2021). In fact, as the dramatic reversal of COVID vaccine attitudes in the Philippines shows, communities in the Global South are arguably more sensitive and averse to being dictated to, either by foreign powers or their own governments. Both China and India have taken a leftist populist approach in their vaccine diplomacy. That is, they project themselves as a champion for the underprivileged in the developing world against aloof and 'greed'-driven Western institutions. They both have demonstrated significant hard power (be it funding for clinical trials or production capacities) in seeking global prominence. Yet beyond oppositional politics they struggle to assure public confidence or exert soft power to appeal to social expectations.

Conclusion

To some extent, the parallel development of China and India is a form of tango: a choreography of an intimate yet flexible embrace of two long-faced dancers, tempo-ed by an incisive and provocative on-again, off-again beat. Since 2014, China has become India's largest trading partner. The transactions between China and India are five times that between China and Pakistan (Shepard, 2017). It is a rivalry with deep-rooted co-dependence. What should we make of their increasing roles internationally? What challenges do they bring forward?

It would be wrong to frame the global implications of the rise of China and India as the *consequence* of their emerging domestic capacities. Rather, the global is very much *embedded* in both China's and India's strategy of national revival. Despite (or because of) the restrained domestic capacity, they are attentive to investing in transnational coalitions. There is an evident leftist populism in both countries' development rationales, pitting the Global South scientific ambition against Global North epistemic hegemony. Through such a lens, it is easier to identify the latent impact of South–South collaboration in terms of norm formation. Although most of the norms Chinese institutions are establishing with BRI partners remain technical and they may not necessarily be non-compatible with Global North preferences, the shifting mindset over who makes the norms and how they are circulated alters underlying global power dynamics.

The leftist populist lens also makes visible why particular aspirations are more motivating than others to China and India, and how the two countries want be seen. For example, why vaccine diplomacy matters more to China and India than the economically more lucrative and politically less controversial option of selling their services to international organisations.

One of the legacies of the pandemic is that science will become a more visible and more critical element in global politics and international relations. Yet the delivery of good science relies not only on hard power, but also on an equivalent capacity to wield soft power. In this sense, populism is not helpful. It thrives on creating political divisions rather than healing them. More importantly, counter-hegemony itself does not constitute a more cosmopolitan world vision nor does it automatically generate power. Both China and India have yet to establish their credibility as a leader in science. The rise of the two Asian countries may well change global dynamics and diversify how science is done, but it may not revolutionise the logic of how science is governed and institutionally applied.

This leads to our last point. To recall the Chinese media's summary of the global vaccine rollout disparity as being 'under two different skies', there is an acute sense of growing inequality and lack of respect and/or recognition for the needs of the Global South. To some extent, India's and China's urge to rise is a manifestation of the challenge of inequality, rather than being the challenge itself. Good science is not populist but it should be people-centred, for it is not primarily intended to subvert but to understand. Science is a project of balancing dissent so as to reach conditional consensus.

Such consensus should not be just among elites or powerful institutions. It has long been argued that 'civic epistemology' is just as important as scientific epistemology, and upgrading it to the 'global' scale remains easier said than done. A proper accommodation of the plurality of civic epistemologies demands not a simple transference of power, but may need a revolution in the logic of how science is governed.

6

What global science will have been

This book addresses a critical problem of the contemporary global governance of the life sciences. Science is at large; it is practised by an ever increasing number of individuals across the globe, not necessarily in conformity or in meaningful dialogue with each other. There has been increasing de-territoriality of science practices, bottom-up brokerage of policies and knowledge, and a multiplicity and fragmented public sphere. As science outgrows conventional colonies of expertise and authorities, it challenges governance norms that have been founded, developed and led by the West.

The rise of China and India poses a magnifying glass effect of these challenges. Thus, our examination of critical events in both countries illuminates how they have become the 'disobedient' new rivals despite (or because) of their continuous exchange and dialogue with the West. This book could be seen as a discussion on global governance by moving from a fixation on the West to an 'off-centred' view of the Global South. We contribute to the ongoing academic effort of 'provincialis[ing] Europe' so as to allow a more accurate vision of global science to emerge (Chakrabarty, 2000). The Western apprehension attached to China's and India's rise lies not so much in their attempts to pose an 'alternative' to Western science. Rather, their rise is seen to be subversive precisely because it presents varieties – such as room for negotiation and co-existence of authorities – within the same scientific paradigm (Mignolo, 2011: xxviii–xxix). But not every path taken by China's and India's sciences are sensible or sustainable. Some, as discussed in the book, are even counterproductive.

More importantly, scientific possibilities have been exploited by private ventures, curious individuals and new amalgams of social interests. For these issues, neither the East nor the West has good answers, and both need a radical rethink of how to deploy their power and resources transnationally. The title of this final chapter expresses the urgency we feel towards such a rethinking. The future anterior framing of 'will have been' draws attention to the fact that the anticipated future is embedded in the present (Barlow, 2004). It also underlines the necessity that to ensure accountable science for the future, we need to start de-colonising the way we think about science governance now.

What does a de-colonised global science governance look like? We don't have the answer and frankly, we are not in a position to offer an answer; for it must be addressed collectively and may have multiple solutions as to how it can be achieved. We do believe, however, in breaking down the 'how' question into the when, where and who. Thus, we share our final thoughts on the topics of *time*, *place* and *people* that our examination has underlined.

Time

Good science is always about being conscious of time. From the controlled cultivation of embryos in a petri dish to the tireless pursuit of timely discoveries, attentiveness to time is how science understands and speaks to social concerns. Time also plays a central role in the grand scheme of the project of modernity. Science pertains to the shaping of our collective future and, especially in the case of the Global South, science is also about lifting a community out of its colonial past. Science is simultaneously about collective dreams and collective memories.

But how we think about time could also be an obstacle that blocks our view from the path to effective global governance. Arguably, since the end of the Cold War, a preoccupation of global governance has been about bringing everyone into 'the same time zone' – in other words, bring developing countries up to speed. Gearing to Western standards, and harmonising global policies – since the Cold War, late developing countries have been preoccupied with making up for 'lost' time, which has led to what sociologist Kyung-Sup Chang (1999) characterised as 'compressed modernity'. Assessment about development has also implicitly been about countries' relative position along the same timeline. This is particularly the case for science governance. It is not uncommon to hear stakeholders from both the Global North and Global South make comments such as 'China is 20 years behind the West' or 'India has another 10-year gap to catch up'. But what if catching up is not about competing along a *linear* scientific progression or a particular ordering of how science should be done? What if catching up is about developing a similar capacity to command science so that science can speak to *their* concerns? Sure, not all concerns and needs are of equal importance. But new governance challenges will not be solved by waiting or urging others to conform. As argued in Chapter 4, a truly global and productive governing approach to the life sciences must escape the view that science is being led by an echelon of (Western) elites; rather, it needs to sync with the actual landscape of who could be doing science and needs to be relevant to these new concerns. Relevance does not mean agreement but it is the base of where disagreement can be meaningfully discussed and negotiated.

Old power imbalances may blind us from recognising real-time change. As discussed in Chapter 2, one reason that gave rise to the Gordian knots of contemporary science governance is a North–South mismatch of perceptions in science delivered by the Global South. In 2012, Joy wrote a book titled *The Cosmopolitanization of Science*. It demonstrated a growing awareness among Chinese scientists and ethicists that there was not a monolithic authority to dictate the 'correct' norm of science; rather, norms are forged through a cosmopolitan negotiation of scientific divergence. The cosmopolitan flows of knowledge and risk in Asia was elegantly demonstrated in Aiwha Ong's (2016) *Fungible Life*. Being epistemically disobedient in the organisation of science, especially in emerging areas, is not a defiance but an inevitable aspect of a polycentric science. As seen in Chapters 3 and 5, the new generation of internationally trained researchers are no longer satisfied with simply 'selling their production capacity' to global science, as shown in the cases of BGI and the COVID vaccine race. They want to be recognised as equal partners. The issue is not that Global South researchers need to be brought into the same time zone but that they're already in the same room. Yet outdated impressions of them being the followers makes it difficult to see they may well be in a different stage of development. We are contemporaneously faced with similar socio-technical uncertainties and an absence of foresight. We highlight the seemingly obvious point on 'contemporaneity', for it is this point that often gets overlooked. Scientific communities, regardless of their position in the North or South respond and (re)act to emerging science not simply due to historical cultural differences but also due to concurrent inspirations and provocations in global exchange. Thus, to ponder the question what the life sciences 'will have been' is also to remind ourselves that the authority of governance is increasingly dependent on its current perceived fairness and inclusivity.

Good science governance is always about being conscious of time. Its effectiveness comes from continuously assessing, comparing, reflecting and mediating a world of scientific divergence. As pointed out by Bijker and his colleagues, the devices of science governance that developed during the era where institutional science was dominant lack the means to address the twenty-first century's pervasiveness of scientific culture (Bijker, Bal and Hendriks, 2009). Our governing outlook needs to sync with the times.

Place

By place, we do not simply refer to the steadily growing place-focused science investment strategies that countries like the UK are championing (McCann, 2019). By place, we highlight that science occupies a political geography. When we think about science governance, especially at the international level, too often we think about 'space', but not 'place' (Gieryn, 2000; Roudometof, 2019). To draw on Chinese-American geographer Yi-Fu Tuan's (1977 [2001]) seminal work, *Space and Place: The Perspective of Experience*, 'space' is of the abstract where as 'place' is of the particular. Space is the environment that generates possibilities, relations and movements. Place is the geographic locality and material uniqueness that 'humanises' space (Tuan, 1977 [2001]: 54). In science, space is where the ideas and the ideals flow, while place is what provides actual empowerment. Space is freedom whereas place gives meaning to that freedom (Tuan, 1977 [2001]: 3).

'I am where I think', to recall Mignolo's (2011: 92–3) insights on agency cited in Chapter 1. It is not simply about the 'situatedness' of an actor, but a case of highlighting that agency itself is an outcome of a matrix of power. To be able to think from and with others, we need to first be able to locate where they are (Mignolo and Walsh, 2018). To acquire a de-colonised vision of science governance, it is time to step out of a transnational sphere of 'space' and walk into the particularities of a 'place'.

National habitus is a useful structure to help us make sense of various places of science. As clarified at the beginning of the book, this is by no means suggesting that everything can be attributed to a monolithic 'national' character. In fact, one of the first things one notices inside a nation-state, such as exhibited by the Bt crops controversies in India, is that even 'state science' does not guarantee national consensus. Similar phenomena are mirrored in how growing scepticism among the Chinese public has subverted state ambitions on measles vaccinations and on rolling out GM crops. But the nation-state remains a practical unit of analysis because it marks the physical territories of policy and funding remits that structures scientific practices. As Chapter 2 explained, the two countries' self-perception of their habitus and their subaltern positions in the field of global science help us to comprehend the general outlook that framed their respective strategies.

Places are maintained by evolving interests. Even in authoritarian countries such as China, the internal dynamic of how science is conceived and delivered has been significantly reshaped by the five sets of relations examined in Chapter 3. More importantly, regardless of whether it is state–scientist, state–ethicist, science–public, science–science or state–science relations, these are formulated in response to a confluence of local and global discourses. This echoes Latour's (1993: 120) insight that it is the 'skein of networks' that ties the technoscientific space into a place.

Places are being made. The examples of South–South platforms, such as the Belt and Road Regional International Standardization Committee discussed in Chapter 5, demonstrates how new geographies of science can be formed on top of old borders.

Finally, place is also a verb. To comprehend the contingency and sometimes self-conflicting science–society relations in Global South societies is to understand where they place science in the dual process of modernisation and globalisation. What lies at the core of protesting against scientific hegemonies is that the development of science lies not only in a matrix of power but in a matrix of values too.

Yi-fu Tuan was of the view that 'when space feels thoroughly familiar to us, it has become place' (Tuan, 1977 [2001]: 73). Place can be 'a favourite armchair, or 'at the other extreme of the whole earth' (Tuan, 1977 [2001]: 149). In some sense, science governance at the global level is about extending visions of science from our respective armchairs to the whole earth. To what extent a vision, along with its ordering of the social, can be anchored at different places depends on how much its advocates can enable others to feel welcomed and comfortable within such spaces.

People

'Science is becoming human' Ulrich Beck (1992: 167) announced. 'It is packed with errors and mistakes … The opposite attracts, it always has opportunities as well. The scene is becoming colorful. If three scientists get together, fifteen opinions clash'. This is not to say that science has sunk to a state of anarchy. This 'greater versatility' (Beck, 1992: 167) observed in contemporary science only makes what anthropologist Michael Fischer (2013b) calls the study of the 'peopling of technologies' ever more important.

'Peopling' includes tracing the individuals that organise and deliver science, knowing who the movers and shakers are, who benefits from science, or is exploited by it, or simply precluded from it. More importantly, it's about understanding what to expect from people's 'unsocial socialities' (Fischer, 2009, 242). In other words, understanding the peopling of science is to understand real-world reflections and negotiations over the question of whom science is for.

Science is never simply an object of policy. It is about people and the power asymmetry between them (Nowotny, 2007; Palladino and Worboys, 1993). To de-colonise science governance away from the ivory tower and a hegemonic view is to have the courage to leave the false reassurance created by a 'bureaucratic amplification of credibility' (Chapter 4). A de-colonised approach to science governance is to recognise that our collective resilience does not hinge so much on an equal distribution of the same set of knowledge but on an ecology in which different knowledge-ways have equal opportunity to contribute to our collective wisdom (Santos, 2014). As such, we need 'a *multitude of policy rooms* distributed (and competing) throughout the science and innovation system' not just within pockets of communities in the West but in the Global South as well (Nowotny, 2007: 428, original emphasis).

The peopling of science itself comes with challenges. At a macro level, 'people' can be instrumentalised to legitimise particular political interests, as we've seen in the left-wing science populism exhibited in both China and India (Chapter 5). The peopling of science in the form of populism may further fragment rather than unite the world. At a micro level, either through radical nation-wide open access, or through aggressive privatisation of translational research, inspiring and empowering people is a necessity but it also necessitates an upgrade for governance.

Hannah Arendt (1958 [1998]: 5) forthrightly said her reflection on the human condition was about one central question: 'what are we doing?' It is long overdue for us to pause and think about this question before setting off further. However, some would argue that with the added pressure coming from the rapid ascent of the Global South, we cannot afford the time to pause, as we must plough ahead. But does the logic that we don't have the time to pause hold true?

We opened our discussion with 'the perfect scandal'. The CRISPR baby case is not a simple case of illegal practice but incorporated a range of issues discussed in this book – issues that were rooted in and further reinforced mutual scepticism between the Global North and the Global South. One thing we haven't fully addressed is the worry of a slippery slope of technological determinism exhibited in this case, both in Jiankui He's confidence of being on the right side of history and in the softening of views of germline editing in Anglo-American circles prior to the scandal. We are in full agreement that every new technology needs to be promoted with caution, that we need to think about what we are doing before committing ourselves to its unknown consequences. But we also think that the root of the problem lies not in an ideological fight against technological determinism but in having the courage to face human nature. As Arendt rightly pointed out, it is unlikely that *homo faber*, man the builder and fabricator, can be completely restricted. For it is the recognition of action as a human prerogative that constitutes the modern mentality (Arendt, 1958 [1998]: 289–304). Thus, while gatekeeping is necessary, it may not be sufficient. We need to move from a governing logic that depends on dictating 'who can do science' to one that anticipates 'who could do science'. If we take science literacy and scientific citizenship with the humility and sincerity needed, then the doing of science and its associated ethics are naturally everybody's business (Franklin, 2019; Jasanoff and Hilton, 2017; Wilsdon and Willis, 2004). This is not to deny the authority nor to degrade the value of professional science or bioethics but to rightly acknowledge the significance science plays in every aspect of all our lives.

Arendt was of the view that in 'an ocean of uncertainty' illuminated by technoscience advancement, we can nevertheless establish 'isolated islands of certainty' by people voluntarily entering into mutual promises and 'act[ing] in concert' (Arendt, 1958 [1998]: 244). Such promises are made 'not by an identical will ... but by an agreed purpose'. And it is the 'force of mutual promise' that can preserve us 'limited independence' from future risks and render us 'the capacity to dispose the future as though it were the present' (Arendt, 1958 [1998]: 245).

This leads back to our first point on 'time'. Science has long ceased to be a globalised venture in the sense of having a monolithic view and uniform practices. But science remains universal, not so much in its practice or norms, but as a universal symbol of hope and a reminder of our interconnectedness, both in the present and in our common future. Perhaps it is high time that we consult each other's views on 'what global science will have been'. After all, it is not the conclusions reached but the questions we ask that make us who we are.

Bibliography

All URLs were live as of April 2021

Abbott, A., and Cyranoski, D. (2001). China plans 'hybrid' embryonic stem cells. *Nature*, 413, 339.

AFP (Agence France-Presse) (2021). Waiver war at WTO over Covid jab intellectual property rights. *The Economic Times*, 1 March. https://economictimes.indiatimes.com/news/international/business/waiver-war-at-wto-over-covid-jab-intellectual-property-rights/articleshow/81272975.cms?from=mdr.

Al Jazeera (2021). Rich nations 'hoarding' a billion doses of excess COVID vaccine. *Al Jazeera*, 19 February. www.aljazeera.com/news/2021/2/19/covid-vaccine.

Alexander, J. C. (2013). *The Dark Side of Modernity*. London: Polity.

Alok S., and Al-Zoubi, Ziad M. (2016). Rethinking on ethics and regulations in cell therapy as part of neurorestoratology. *Journal of Neurorestoratology*, 6, 1–14.

Anderson, W. (2002). Introduction: postcolonial technoscience. *Social Studies of Science*, 32, 643–58.

Anderson, W., and Adams, V. (2008). Pramoedya's chickens: postcolonial studies of technoscience. In E. J. Hackett et al. (eds), *The Handbook of Science and Technology Studies*. Cambridge, MA: MIT Press.

Appiah, K. A. (2006). *Cosmopolitanism: Ethics in a World of Strangers*. London: Penguin.

Arendt, H. (1958 [1998]). *The Human Condition*. 2nd edn. Chicago: University of Chicago Press.

Arnold, D. (1993). *Colonizing the Body: State Medicine and Epidemic Disease in Nineteenth-Century India*. Berkeley, CA: University of California Press.

Arora, N., and Das, K. N. (2021). Pfizer wants to make Covid vaccine in India if faster clearance, export freedom assured. *Reuters*, 10 March. www.reuters.com/article/us-health-coronavirus-india-pfizer/exclusive-pfizer-wants-to-make-vaccine-in-india-if-faster-clearance-export-freedom-assured-sources-idUKKBN2B21AY.

Aslanidis, P. (2018). Populism as a collective action master frame for transnational mobilization. *Sociological Forum*, 33, 443–64.

Au, L., and da Silva, R. G. L. (2021). Globalizing the scientific bandwagon: trajectories of precision medicine in China and Brazil. *Science, Technology & Human Values*, 46, 192–225.

Bak, H. (2018). Research misconduct in East Asia's research environments. *East Asian Science, Technology and Society*, 12(2), 117–22.

Bala, P. (2012). *Contesting Colonial Authority: Medicine and Indigenous Responses in Nineteeth- and Twentieth-century India*. Lexington, KY: Lexington Books.

Balmer, A. S., et al. (2015). Taking roles in interdisciplinary collaborations: reflections on working in post-ELSI spaces in the UK synthetic biology community. *Science & Technology Studies*, 28(3), 3–25.

Baltimore, D., et al. (2018). Statement by the Organizing Committee of the Second International Summit on Human Genome Editing. National Academies of Sciences, Engineering, and Medicine, US, 28 November. www.nationalacademies.org/news/2018/11/statement-by-the-organizing-committee-of-the-second-international-summit-on-human-genome-editing.

Barad, K. (2007). *Meeting the Universe Halfway: Quantum Physics and the Entanglement of Matter and Meaning*. Durham, NC: Duke University Press.

Barlow, T. E. (2004). *The Question of Women in Chinese Feminism*. Durham, NC: Duke University Press.

Barr, R. R. (2009). Populists, outsiders and anti-establishment politics. *Party Politics*, 15, 29–48.

BBC (1996). First GM food goes on sale in UK, 5 February. http://news.bbc.co.uk/onthisday/hi/dates/stories/february/5/newsid_4647000/4647390.stm.

BBC (2019). China jails 'gene-edited babies' scientist for three years, 30 December. www.bbc.co.uk/news/world-asia-china-50944461.

BBC (2021). 'Greed' and 'capitalism' helped UK's vaccines success, says PM, 25 March. www.bbc.co.uk/news/uk-politics-56504546.

Beaumont, P. (2021). China stalls WHO mission to investigate origins of coronavirus. *Guardian*, 6 January. www.theguardian.com/world/2021/jan/06/china-stalls-who-mission-to-investigate-origins-of-coronavirus.

Beck, U. (1992). *Risk Society: Towards a New Modernity*. London: Sage.

Begley, S. (2018). He took a crash course in bioethics. Then he created CRISPR babies. *Statnews*, 27 November. www.statnews.com/2018/11/27/crispr-babies-creator-soaked-up-bioethics/.

Beijing Youth Daily (2016). The public-benefit nature of public hospitals and medical colleges cannot be cancelled. *Beijing Youth Daily*, 27 July. http://news.ifeng.com/a/20160727/49667664_0.shtml.

BGI (Beijing Genomic Institute) (2020). Research output. *Beijing Genomic Institute*, 29 December. www.genomics.cn/result.html.

Bell, S., and Hindmoor, A. (2009). *Rethinking Governance: The Centrality of the State in Modern Society*. Cambridge: Cambridge University Press.

Bera, S., and Sen, S. (2016). Govt cuts Bt cotton royalty fees by 74%. *Mint*, 10 March. www.livemint.com/Politics/NdDYRxsayfh2655qOqy7mI/Centre-notifies-Bt-cotton-seed-prices-slashes-royalty-fees.html.

Bhagwat, R. (2012). Mahyco banned from selling Bt cotton seeds in Maharashtra. *Times of India*, 10 August. http://timesofindia.indiatimes.com/articleshow/15427722.cms?utm_source=contentofinterest&utm_medium=text&utm_campaign=cppst.

Bhambra, G. K. (2014). Introduction: knowledge production in global context: power and coloniality. *Current Sociology*, 62(4), 451–6.

Bharadwaj, A. (2012). Enculturating cells: the anthropology, substance, and science of stem cells. *Annual Review of Anthropology*, 41, 303–17.

Bharadwaj, A. (2013). Ethic of consensibility, subaltern ethicality: the clinical application of embryonic stem cells in India. *BioSocieties*, 8, 25–40.

Bharadwaj, A. (2014). Experimental subjectification: the pursuit of human embryonic stem cells in India. *Ethnos*, 79(1), 84–107.

Bharadwaj, A. (2016). The Indian IVF saga: a contested history. *Reproductive Biomedicine and Society Online*, 2, 54–61.

Bharadwaj, A., and Glasner, P. (2009). *Local Cells, Global Science: The Rise of Embryonic Stem Cell Research in India*. London: Routledge.

Bhidé, A. (2008). *The Venturesome Economy: How Innovation Sustains Prosperity in a More Connected World*. Princeton, NJ: Princeton University Press.

Bijker, W. E., Bal, R., and Hendriks, R. (2009). *The Paradox of Scientific Authority: The Role of Scientific Advice in Democracies*. Cambridge, MA: MIT Press.

Biotech China (2021). Global times: WHO expert team visits Wuhan Virology Institute: smearing due to inability to accept the truth, *Biotech-China*, n.d. https://mp.weixin.qq.com/s/KCcGtWbb5fkcdihLaDt-6g.

Biotechnology Research Institute of CAAS (Chinese Academy of Agricultural Sciences) (2010). Phytase transgenic maize approval. *CAAS*, 23 November. http://bri.caas.cn/cgzh/zhxm/76512.htm.

Birch, K., and Tyfield, D. (2012). Theorizing the bioeconomy: bivalve, biocapital, bioeconomics or … what? *Science, Technology & Human Values*, 38, 299–327.

Blakely, R. (2009). Doctor's stem-cell 'miracles' raise medical doubts; patients eager to pay for treatment banned in Britain. *The Times* (London), 7 November. www.thetimes.co.uk/article/stem-cell-experts-sceptical-of-dr-geeta-shroffs-miracle-cure-claims-36hlct67bcb.

Bloomfield, B. P. and Vurdubakis, T. (1995). Disrupted boundaries: new reproductive technologies and the language of anxiety and expectation. *Social Studies of Science*, 25, 533–51.

Bogner, A., and Menz, W. (2010). How politics deals with expert dissent: the case of ethics councils. *Science Technology & Human Values*, 35, 888–914.

Bourdieu, P. (1969). Intellectual field and creative project. *Social Science Information*, 8, 89–119.

Bourdieu, P. (1990). *The Logic of Practice*. Cambridge: Polity.

Bourdieu, P., and Wacquant. L. (1989). Towards a reflexive sociology: a workshop with Pierre Bourdieu. *Sociological Theory*, 7(1), 26–63.

Bourdieu, P., and Wacquant, L. (1992). *An Invitation to Reflexive Sociology*. Cambridge: Polity.

BRAI Bill (2013). The Biotechnology Regulatory Authority of India Bill. Bill No. 57 of 2013. *As introduced in Lok Sabha* (The Lower House of the Parliament of India).

Braude, P., Minger, S. L., and Warwick, R. (2005). Stem cell therapy: hope or hype? *British Medical Journal*, 330.

Bray, F. (2008). Science, technique, technology: passages between matter and knowledge in imperial Chinese agriculture. *British Journal for the History of Science*, 41, 319–44.

Brenner, L. H., and Bal, B. S. (2012). Regulation or innovation: United States v regenerative sciences. *Orthopedics Today*, 10 October. www.healio.com/news/orthopedics/20121010/10_3928_1081_597x_19300101_02.

Brubaker, R. (2017). Why populism? *Theory and Society*, 46, 357–85.

Business Standard (2008). Stem cell research making progress in India. *Business Standard*, 14 June. www.business-standard.com/article/economy-policy/stem-cell-research-making-progress-in-india-108021901066_1.html.

Butt, M., Mohammed, R., Butt, E., Butt, S., and Xiang, J. (2020). Why have immunization efforts in Pakistan failed to achieve global standards of vaccination uptake and infectious disease control? *Risk Management and Healthcare Policy*, 13, 111–24.

Cai, H., and Wang, Y. (2020). Success both in rescue and scientific research: comparison of medical papers published by Chinese in two epidemics. *Adverse Drug Reactions Journal*, 22(6), 329–32.

Caiani, M., and Padoan, E. (2020). Setting the scene: filling the gaps in populism studies. *Partecipazione e Conflitto*, 13, 1–28.

Cao, C. (2004). *China's Scientific Elite*. Abingdon: Routledge.

Cao, C., Li, N., Li, X., and Liu, L. (2018). Reform of China's science and technology system in the Xi Jinping Era. *China: An International Journal*, August, 120–41.

Cao, C., and Suttmeier, R. P. (2001). China's new scientific elite: distinguished young scientists, the research environment and hopes for Chinese science. *China Quarterly*, 168, 960–84.

Carlson, R., Reiter, D., and Lu, K. (2018). Ebola vaccine development race between the USA and China. *Precision Vaccines*, 31 March. www.precisionvaccinations.com/ebola-vaccine-candidate-v920-approaching-finish-line#medical-review.

CAS (Chinese Academy of Science) (2004). Chen Zhu et al. call for enhanced bioethics research. Chinese Academy of Sciences. www.cas.cn/xw/zjsd/200401/t20040108_1684473.shtml.

CAS (Chinese Academy of Sciences) (2009) The Chronicles of the Chinese Academy of Sciences: 1998. Chinese Academy of Sciences. www.cas.cn/jzzky/ysss/bns/200909/t20090928_2529283.shtml.

CAS (Chinese Academy of Science) (2019). Chinese Sciences Academy Provides 268 million Dollars for BRI Projects. Chinese Academy of Sciences, Beijing, 19 April. http://english.cas.cn/newsroom/news/201904/t20190419_208623.shtml.

CASTED (Chinese Academy of Science and Technology for Development) (2019). Second China-India science and innovation collaboration research symposium opens in China. *China Association for International Science and Technology Cooperation*. www.caistc.com/a/member/huiyuanzhijia/2020/0119/458.html.

Centeno, C. J., et al. (2016). A multi-center analysis of adverse events among two thousand, three hundred and seventy two adult patients undergoing adult autologous stem cell therapy for orthopaedic conditions. *International Orthopaedics*, 1–11.

Center for Genetics and Society (2015). Public Interest Group Calls for Strengthening Global Policies Against Human Germline Modification. Center for Genetics and Society press statement. 22 April.

Chadwick, A. (2006). *Internet Politics: States, Citizens, and New Communication Technologies*. New York: Oxford University Press.

Chakrabarty, D. (2000). *Provincializing Europe: Postcolonial Thought and Historical Difference*. Princeton, NJ: Princeton University Press.

Chambers, D. W., and Gillespie, R. (2000). Locality in the history of science: colonial science, technoscience, and indigenous knowledge. *Osiris*, 15, 221–40.

Chandrashekar, S. (1995) Technology priorities for India's development: need for restructuring. *Economic and Political Weekly*, 30, 2739–48.

Chang, K. (1999). Compressed modernity and its discontents: South Korean society in transition. *Economy and Society*, 28, 30–5.

Chattopadhyay, S. (2012). Guinea pigs in human form: clinical trials in unethical settings. *Lancet*, 379, e53.

Chaudhuri, S. (1984). Manufacturing drugs without TNCs: status of indigenous sector in India. *Economic and Political Weekly*, XIX(31–33), 1341–84.

Chauvin, L. O., Faiola, A., and Dou, E. (2021). Squeezed out of the race for Western vaccines, developing countries turn to China. *Washington Post*, 16 February. www.washingtonpost.com/world/2021/02/16/coronavirus-peru-china-vaccine-sinopharm-sinovac/.

Chee, L. P. Y., and Clancey, G. (2013). The human proteome and the Chinese liver. *Science, Technology & Society*, 18, 307–19.

Chen, D. (2015). To publish or not publish: the case with the paper gene modified embryos. *Science and Technology Daily* (Chinese), 30 April. http://scitech.people.com.cn/n/2015/0430/c1057-26927951.html.

Chen, F., Wu, C., and Yang, W. (2016). A new approach for the cooperation between academia and industry: an empirical analysis of the triple helix in east China. *Science, Technology & Society*, 21, 181–204.

Chen, H. (2013). Governing international biobank collaboration: a case study of China Kadoorie biobank. *Science, Technology & Society*, 18, 321–38.

Chen, L., et al. (2019). Summary report of the 11th Annual Conference of International Association of Neurorestoratology (IANR). *Journal of Neurorestoratology*, 7, 1–7.

Chen, Y., et al. (2003). Embryonic stem cells generated by nuclear transfer of human somatic nuclei into rabbit oocytes. *Cell Research*, 13, 251–64.

Cheng, L-Z., et al. (2006). Ethics: China already has clear stem-cell guidelines. *Nature*, 440, 992.

Cheng, Y. (2018). China will always be bad at bioethics. *Foreign Policy*, 13 April. https://foreignpolicy.com/2018/04/13/china-will-always-be-bad-at-bioethics/.

Chinese Academy of Science (2006). Review of the Chinese Academy of Science's knowledge innovation pilot projects. *China Central Television*, 17 March. www.cctv.com/science/special/C15426/20060317/102120.shtml.

China-South Asia Technology Transfer Center (2020). The Fourth India-China Technology Transfer, Collaborative Innovation Investment Conference a Success. *China-South Asia Technology Transfer Center*, 25 November. http://in.china-embassy.org/chn/zywl/t1834350.htm.

China Business Weekly (2020). Why overseas clinical trials for domestic COVID vaccines? CCTV, 16 September. https://news.cctv.com/2020/09/16/ARTIuZk0SqI9QCnGZ2RuX20T200916.shtml.

Chuan J. (2016). The Wei Zexi case: who is supervising the people's military hospitals? BBC, 16 May. www.bbc.com/zhongwen/simp/china/2016/05/160504_china_military_hostpital_scandal.

Cingta (2018). Post-80er female scientists: youngest shortlisted for changing scholar of the year. *Cingta*, 16 July. www.cingta.com/detail/5914.

Coalition for a GM-Free India. (2011). Biotechnology Regulatory Authority of India Bill, 2011 (BRAI): Wrong Bill by the wrong people, for the wrong reasons – a critique by the Coalition for a GM-Free India. *Biotechnology* (August), 1–11. http://indiagminfo.org/wp-content/uploads/2011/09/brai-2013-coalition-critique.pdf.

Cohen, J. (2019a). The untold story of the 'circle of trust' behind the world's first gene-edited babies. *Science*, 1 August. www.sciencemag.org/news/2019/08/untold-story-circle-trust-behind-world-s-first-gene-edited-babies.

Cohen, J. (2019b). Embattled Russian scientist sharpens plans to create gene-edited babies. *Science*, 369, 1435. DOI: 10.1126/science.aaz9337.

Collier, S., and Ong, A. (2005). Global assemblages, anthropological problems. In A. Ong and S. Collier (eds), *Global Assemblages: Technology, Politics, and Ethics as Anthropological Problems*. Malden, MA: Blackwell.

Comaroff, J., and Comaroff, J. (1992). *Ethnography and the Historical Imagination*. Boulder, CO: Westview Press.

Crane, J. (2010). Adverse events and placebo effects: African scientists, HIV, and ethics in the 'Global Health Sciences'. *Social Studies of Science*, 40(6), 843–70.

Crombie, A. C. (1990). *Science, Optics, and Music in Medieval and Early Modern Thought*. London: Hambledon Press.

Cui, J., Li, F., and Shi, Z-L. (2019). Origin and evolution of pathogenic coronaviruses. *Nature Reviews Microbiology*, 17, 181–92.

Curwen, L. (2020). The risks behind the hype of stem-cell treatments. BBC News, 7 January. www.bbc.co.uk/news/health-51006333.

Cyranoski, D. (2005). Fetal-cell therapy: paper chase. *Nature*, 437, 810–11.

Cyranoski, D. (2010). FDA challenges stem cell clinic. *Nature*, 466, 909.

Cyranoski, D. (2018a). CRISPR-baby scientist fails to satisfy critics. *Nature*, 564, 13–14.

Cyranoski, D. (2018b). China creates grand science ministry. *Nature*, 555, 425–6.

Cyranoski, D. (2019). The CRISPR-baby scandal: what's next for human gene-editing? *Nature*, 566, 440–2.

Cyranoski, D., and Ledford, H. (2018). Genome-edited baby claim provokes international outcry. *Nature*, 563, 607–8.

Cyranoski, D., and Reardon, S. (2015). Embryo editing sparks epic debate. *Nature*, 520, 593–4.

DBT (Department of Biotechnology) (2017). *Introduction to the Department of Biotechnology*. Ministry of Science and Technology. https://dbtindia.gov.in/about-us/introduction.

DBT-ICMR (2007). *Guidelines for Stem Cell Research and Therapy*.

Das, V. (1996). *Critical Events: An Anthropological Perspective on Contemporary India*. Oxford: Oxford University Press.

Datta, S. (2018a). An endogenous explanation of growth: direct-to-consumer stem cell therapies in China, India and USA. *Regenerative Medicine*, 13(5), 559–79.

Datta, S. (2018b). Dynamics of evidence and trust in user-user engagement: the case of experimental stem cell therapies. *Critical Public Health*, 28(3), 352–62.

Datta Burton, S. (2018). Policy and practice: parallel convergence and experimental stem cell therapies. Case study: India and China. PhD dissertation. Faculty of Social Science & Public Policy, King's College London.

Datta Burton, S., et al. (2021). Rethinking value construction in biomedicine and healthcare. *Biosocieties*, 1–24.

Dear, P. (2001). Science studies as epistemography. In J. Labinger and H. Collins (eds), *The One Culture? A Conversation about Science*. Chicago: University of Chicago Press.

Deccan Herald. (2008). Experts seek stem cell regulations. *The Deccan Herald*, 19 February. http://archive.deccanherald.com/deccanherald.com/content/Feb192008/state2008021953056.asp.

De Cleen, B., and Stavrakakis, Y. (2017). Distinctions and articulations: a discourse theoretical framework for the study of populism and nationalism. *Javnost – The Public*, 24(4), 301–19.

De Cleen, B., et al. (2017). Constructing the 'refugee crisis' in Flanders: continuities and adaptations of discourses on asylum and migration. In M. Barlai, B. Fähnrich, C. Griessler, and M. Rhomberg (eds), *The Migrant Crisis: European Perspectives and National Discourses*. Berlin: LIT Verlag.

De Cleen, B., Moffitt, B., Panayotu, P., and Stavrakakis, Y. (2020). The potentials and difficulties of transnational populism: the case of the Democracy in Europe Movement 2025 (DiEM25). *Political Studies*, 68, 146–66.

Deng, C. (2020). China seeks to use access to Covid-19 vaccines for diplomacy. *Wall Street Journal*, 17 August. www.wsj.com/articles/china-seeks-to-use-access-to-covid-19-vaccines-for-diplomacy-11597690215.

Deng, X. (1993). *Selected Works of Deng Xiaoping*, Volume 3. Beijing: People's Publishing House.

Dennis, C. (2002). China: stem cells rise in the East. *Nature*, 419, 334–6.

Department of Science and Technology, India (2020). Science, Technology and Innovation Policy (Draft STIP Doc 1.4). *Government of India*, December. https://dst.gov.in/sites/default/files/STIP_Doc_1.4_Dec2020.pdf.

Desiraju, G. R. (2012). Bold strategies for Indian science. *Nature*, 484, 159–60.

Dickson, D. (1984). *The New Politics of Science*, Chicago: University of Chicago Press.

Dickenson, D., and Darnovsky, M. (2019). Did a permissive scientific culture encourage the 'CRISPR babies experiment? *Nature Biotechnology*, 37, 355–7.

DiEM25 (2016). A Manifesto for Democratising Europe (Long Version). DiEM25. February. https://diem25.org/wp-content/uploads/2016/02/diem25_english_long.pdf.

Dobkin, B. H., Curt, A., and Guest, J. (2006). Cellular transplants in China: observational study from the largest human experiment in chronic spinal cord injury. *Neurorehabil Neural Repair*, 20(1), 5–13.

Doering, O. (2004). Chinese researchers promote biomedical regulations: what are the motives of the biopolitical dawn in China and where are they heading? *Kennedy Institute of Ethics Journal*, 14(1), 39–46.

Dube, R., and Magalhaes, L. (2021). For Covid-19 vaccines, Latin America turns to China and Russia, *Wall Street Journal*, 24 February. www.wsj.com/articles/for-covid-19-vaccines-latin-america-turns-to-china-and-russia-11614186599.

Duggan, M., Garthwaite, C., and Goyal, A. (2016). The market impacts of pharmaceutical product patents in developing countries: evidence from India. *The American Economic Review*, 106(1), 99–135.

Dunford, M., and Qi, B. (2020). Global reset: COVID-19, systemic rivalry and the global order. *Research in Globalization*, 2(100021).

DW News (2021). Rich countries block India, South Africa's bid to ban COVID vaccine patents. DW News. www.dw.com/en/rich-countries-block-india-south-africas-bid-to-ban-covid-vaccine-patents/a-56460175.

Dzau, V. J., McNutt, M., and Bai, C. (2018). Wake-up call from Hong Kong. *Science*, 362, 1215.

The Economist (2019). Red moon rising. *The Economist*, 430(9125), 11. www.economist.com/leaders/2019/01/12/how-china-could-dominate-science.

The Economist (2021). Who gets the jab? *The Economist*, 9 January, 10.

Einsiedel, E. F., and Adamson, H. (2012). Stem cell tourism and future stem cell tourists: policy and ethical implications. *Developing World Bioethics*, 12, 35–44.

Else, H., and Van Noorden, R. (2021). The fight against fake-paper factories that churn out sham science. *Nature*. www.nature.com/articles/d41586-021-00733-5.

Embassy of PR China, India (2020). Media Celebration of the 70th Anniversary of China–India Diplomatic Relations. *Embassy of PR China, India*, 1 April. http://in.chineseembassy.org/chn/dsxx/dshdjjh/t1764667.htm.

Enserink, M. (2006). Selling the stem cell dream. *Science*, 313, 160–3.

Enserink, M. (2013). Golden rice not so golden for Tufts. *Science*, 18 September. www.sciencemag.org/news/2013/09/golden-rice-not-so-golden-tufts.

Estrada, M. S. (2017). Exploring tensions in knowledge networks: convergences and divergences from social capital, actor-network theory and sociologies of the south. *Current Sociology Review*, 65, 886–908.

Ethics Committee of the Chinese National Human Genome Centre at Shanghai (CHGC) (2001). Ethical Guideline on Human Embryonic Stem Cell Research (Recommended Draft). *Chinese Medical Ethics (Zhongguo Yixue Lunlixue)*, 6, 8–9.

EuroStemCell (2016). ISSCR releases new global guidelines for stem cell research. www.eurostemcell.org/isscr-releases-new-global-guidelines-stem-cell-research.

Fan, Y., Zhao, K., Shi, Z-L., and Zhou, P. (2019). Bat coronaviruses in China. *Viruses*, 11(3) (210).

Fanelli, D. (2009). How many scientists fabricate and falsify research? A systematic review and meta-analysis of survey data. *PLOS ONE*, 4(5), e5738.

Fearnley, L. (2020). *Virulent Zones: Animal Disease and Global Health at China's Pandemic Epicenter*. Durham, NC: Duke University Press.

Fejerskov, A. M. (2017). The new technopolitics of development and the Global South as a laboratory of technological experimentation. *Science, Technology & Human Values*, 42, 947–68.

Feng, Y., et al. (2014). Research on long mid-term science, technology and innovation policy of emerging economies: taking India as an example. *China Soft Science*, 2014(9), 182–92.

Financial Times (2006). Timeline: the EU's unofficial GMO moratorium. 19 July. www.ft.com/content/624a88c6–97db-11da-816b-0000779e2340.

Fischer, M. M. J. (2009). *Anthropological Futures*. Durham, NC: Duke University Press.

Fischer, M. M. J. (2013a). Biopolis: Asian science in the global circuitry. *Science, Technology & Society*, 18, 379–404.

Fischer, M. M. J. (2013b). The peopling of technologies. In J. Biehl and A. Petryna (eds), *When People Come First: Critical Studies in Global Health*. Princeton NJ: Princeton University Press.

Fischer, M. M. J. (2018). A tale of two genome institutes: qualitative networks, charismatic voice, and R&D strategies – juxtaposing GIS Biopolis and BGI. *Science, Technology & Society*, 23, 271–88.

Flemes, D. (2009). India-Brazil-South Africa (IBSA) in the new global order: interests, strategies and values of the emerging coalition. *International Studies*, 46(4), 401–21.

Foster, P., and Fleming, N. (2007). Delhi stem cell jabs 'help woman walk again'. *The Telegraph* (UK), 14 April. www.telegraph.co.uk/news/worldnews/1548589/Delhi-stem-cell-jabs-help-woman-walk-again.html.

Fox, C. (2007). *Cell of Cells: The Global Race to Capture and Control the Stem Cell.* New York: W. W. Norton.

Franklin, S. (2019). Ethical research: the long and bumpy road from shirked to shared. *Nature,* 574, 627–30.

Fricker, M. (2007). *Epistemic Injustice, Power and the Ethics of Knowing.* Oxford: Oxford University Press.

Fu, L. (2016). Evident drop in public acceptance of GMOs. *Science and Technology Daily,* 16 May. http://digitalpaper.stdaily.com.

Fujimura, J. H. (1988). The molecular biological bandwagon in cancer research: where social worlds meet. *Social Problems,* 35(3), 261–83.

Gan, N., Hu, C., and Watson, I. (2020). Beijing tightens grip over coronavirus research, amid US-China row on virus origin. *CNN,* 16 April. https://edition.cnn.com/2020/04/12/asia/china-coronavirus-research-restrictions-intl-hnk/index.html.

Gan, X. (2017). The birth of the world's first Ebola vaccine. ScienceNet.cn, 29 October. http://news.sciencenet.cn/htmlnews/2017/10/392365.shtm.

Gao, X. (2013). Research on Deng Xiaoping's 'debate' and 'no debate' directives. *Contemporary China History Studies,* 1, 66–70.

Giddens, A. (1999) Risk and responsibility. *The Modern Law Review,* 62, 1–10.

Gieryn, T. F. (1983). Boundary-work and the demarcation of science from non-science: strains and interests in professional ideologies of scientists. *American Sociological Review,* 48, 781–95.

Gieryn, T. F. (2000). A space for place in sociology. *Annual Review of Sociology,* 26, 463–93.

Giles, J. (2006). Rules tighten for stem-cell studies. *Nature,* 440(7080), 9.

Glover, D. (2002). Transnational corporate science and regulation of biotechnology. *Economic and Political Weekly* (India), 37(27), 2734–40.

Gluckman, P., and Wilsdon, J. (2016). From paradox to principles: where next for scientific advice to governments? *Palgrave Communications,* 2, Article number: 16077.

Go, J. (2013) Decolonizing Bourdieu: colonial and postcolonial theory in Pierre Bourdieu's early work. *Sociological Theory,* 31, 49–74.

Goldenberg, M. J. (2006). On evidence and evidence-based medicine: lessons from the philosophy of science. *Social Science & Medicine,* 62(11), 2621–32.

Gorman, M. (1988). Introduction of Western science into colonial India: role of the Calcutta Medical College. *Proceedings of the American Philosophical Society,* 132(3), 276–98.

Government of India (1958). Scientific Policy Resolution New Delhi. https://indiabioscience.org/media/articles/SPR-1958.pdf.

Government of India (1983). Technology Policy Statement. https://indiabioscience.org/media/articles/TPS-1983.pdf.

Government of India (2003). Science and Technology Policy 2003. https://indiabioscience.org/media/articles/STP-2003.pdf.

Government of India (2020). Building Aatmanirbhar Bharat and overcoming COVID-19. www.india.gov.in/spotlight/building-atmanirbhar-bharat-overcoming-covid-19.

Great Britain and Warnock, M. (1984). *Report of the Committee of Inquiry into Human Fertilisation and Embryology*. London: HMSO.

Greene, W. (2007). *The Emergence of India's Pharmaceutical Industry and Implications for the U.S. Generic Drug Market*. Office of Economic Working Paper, US International Reade Commission, 05(A).

Grush, L. (2016). UK regulator allows gene editing of human embryos. *The Verge*, 1 February. www.theverge.com/2016/2/1/10885282/gene-edit-dna-crispr-embryo-approval-uk.

Gu, S. (2001). Science and technology policy for development: China's experience in the second half of the twentieth century. *Science, Technology & Society*, 6, 203–35.

Gu, S., et al. (2009). China's system and vision of innovation: an analysis in relation to the strategic adjustment and the medium- to long-term S&T development plan (2006–2020). *Industry & Innovation*, 16(4), 369–88.

Guest, J., Herrera, L. P., and Qian T. (2006). Rapid recovery of segmental neurological function in a tetraplegic patient following transplantation of fetal olfactory bulb-derived cells. *Spinal Cord*, 44, 135–42.

Guo, M. (1978). The spring of science: closing speech at the National Science Conference. *People's Daily*, 1 April. http://scitech.people.com.cn/GB/25509/56813/57267/57268/4001597.html.

Guo, S-W. (2013). China's 'gene war of the century' and its Aftermath: the contest goes on. *Minerva*, 51, 485–512.

Gupta, A. (2011). An evolving science-society contract in India: the search for legitimacy in anticipatory risk governance. *Food Policy*, 36, 736–41.

Gupta, N., and Sharma, A. K. (2002). Women academic scientists in India. *Social Studies of Science*, 32, 901–15.

Hage, J., and Hollingsworth, R. (2000). A strategy for the analysis of idea innovations networks and institutions. *Organizations Studies*, 21, 971–1004.

Hamilton, Jennifer A., et al. (2017). What Indians and Indians can teach us about colonisation: feminist science and technology studies, epistemological imperialism and the politics of difference. *Feminist Studies*, 43, 612–23.

Han, J., and Li, Z. (2018). How metrics-based academic evaluation could systematically induce academic misconduct: a case study. *East Asian Science, Technology and Society*, 12, 165–80.

Harding, S. (1994). Is science multicultural? Challenges, resources, opportunities. *Configuration*, 12, 301–30.

Harding, S. (1998). Is science multicultural? Postcolonialisms, feminisms, and epistemologies. Bloomington, IN: Indiana University Press.

Harding, S. (2001). Multiculturalism and post colonialism: what difference do they make to Western scientific epistemology? *Science Studies*, 14, 45–54.

Harding, S. (2008). *Sciences from Below: Feminisms, Postcolonialities, and Modernities*. Durham, NC: Duke University Press.

Harding, S. (2016). Latin American decolonial social studies of scientific knowledge: alliances and tensions. *Science, Technology & Human Values*, 41, 1063–87.

Harding, S. (2019). State of the field: Latin American decolonial philosophies of science. *Studies in History and Philosophy of Science*, 78, 48–63.

Harris, S. J. (2005). Jesuit scientific activity in the overseas missions, 1540–1773. *Isis*, 96(1), 71–9.

Harrison, M. (1994). *Public Health in British India: Anglo-Indian Preventive Medicine 1859–1914*, Cambridge: Cambridge University Press.

Hasan, Z. (2006). Bridging a growing divide? The Indian National Congress and Indian democracy. *Contemporary South Asia*, 15(4), 473–88.

He, G., et al. (2012). Scientist's public image: its evolution and current state. In W. Li, X. Xu, Y. Zhang, and G. Chen (eds), *2013 Society of China Analysis and Forecast*. Beijing: Social Sciences Academic Press.

Hernandez, J. C., and Gorman, J. (2021). On W.H.O. trip, China refused to hand over important data. *New York Times*, 12 February. www.nytimes.com/2021/02/12/world/asia/china-world-health-organization-coronavirus.html.

Herring, R. J. (2015). State science, risk and agricultural biotechnology: Bt cotton to Bt brinjal in India. *Journal of Peasant Studies*, 42, 159–86.

Hess, D. J. (2007). *Alternative Pathways in Science and Industry*. Cambridge, MA: MIT Press.

Hexun (2017). BGI: unexpected victory of a group of 'science bandits'. *BGI*, 8 July. http://news.hexun.com/2017-07-08/189956969.html.

HFEA (Human Fertilisation and Embryology Authority) (2004) Corporate Plan 2004–2009. London: HFEA.

HFEA (Human Fertilisation and Embryology Authority) (2007). HFEA statement on its decision regarding hybrid embryos. London: HFEA, 5 September. www.hfea.gov.uk/455.html.

HHS (Department of Health and Human Services) (1979). Office of the Secretary of Ethical Principles and Guidelines for the Protection of Human Subjects of Research. The National Commission for the Protection of Human Subjects of Biomedical and Behavioral Research. Department of Health, Education, and Welfare, United States of America.

The Hindu (2010). Moratorium on Bt brinjal. *The Hindu*. 9 February. www.thehindu.com/news/national/Moratorium-on-Bt-brinjal/article 16813609.ece.

The Hindu (2013). Govt investing heavily in science, R&D: Kasturirangan. *The Hindu*. 8 October. www.thehindubusinessline.com/news/national/ govt-investing-heavily-in-science-rd-kasturirangan/article5214504. ece.

The Hindu Businessline (2013). No BRAI Bill, please. *The Hindu Businessline*. 23 August. www.thehindubusinessline.com/opinion/No-BRAI-Bill-please/ article20651786.ece.

Ho, C. M. (2006). Biopiracy and beyond: a consideration of socio-cultural conflicts with global patent policies. *University of Michigan Journal of Law Reform*, 39, 433–542.

Hochschild, A. R. (2016). *Strangers in Their Own Land: Anger and Mourning on the American Right*. New York: New Press.

Holliday, S. J (2016). The legacy of subalternity and Gramsci's national-popular: populist discourse in the case of the Islamic Republic of Iran. *Third World Quarterly*, 37(5), 917–33.

Honneth, A. (1996). *The Struggle for Recognition: The Moral Grammar of Social Conflicts*. Cambridge, MA: MIT Press.

Hu, H., Guo, Y., and Feng, Y. (2016). Research on India's R&D human resource training and attraction policies. *Global Science, Technology and Economy Outlook* (in Chinese), 31, 57–65.

Huang, H. (2010). Neurorestoratology, a distinct discipline and a new era. *Cell Transplantation*, 19, 129–31.

Huang, H., et al. (2018). 2017 yearbook of neurorestoratology. *Journal of Neurorestoratology*, 6, 67–70.

Huang, H., et al. (2019). Clinical neurorestorative therapeutic guidelines for spinal cord injury (IANR/CANR version 2019). *Journal of Orthopaedic Translation*, 20, 14–24.

Huang, J., Qiu, H., Bai, J., and Pray, C. (2006). Chinese urban consumers' knowledge, receptiveness and purchase preference of GM food. *China Soft Science*, 2, 61–7.

Huang, Y., and Li, C. (2012). The 'ethical review' chaos behind the Golden Rice controversy. BJ News, 17 September. http://epaper.bjnews.com.cn/ html/2012-09/17/content_373131.htm.

Hui, M. (2021). China's vaccine diplomacy has an aggressive anti-vax element. *Quartz*, 21 January. https://qz.com/1959855/chinas-coronavirus-vaccine-diplomacy-is-anti-vax/.

Human Fertilisation and Embryology Act (1990). UK Legislation, Chapter 37 (Passed 1 November 1990). www.legislation.gov.uk/ukpga/1990/37/ contents.

HTA (Human Tissue Authority) UK (2004). Human Tissue Act 2004, London: HTA.

Hurlbut, J. B. (2015). Remembering the future: science, law, and the legacy of Asilomar. In S. Jasanoff, and S.-H. Kim, (eds), *Dreamscapes of Modernity: Sociotechnical Imaginaries and the Fabrication of Power*. Chicago, IL: University of Chicago Press.

Hurlbut, J. B. (2018). CRISPR babies raise an uncomfortable reality – abiding by scientific standards doesn't guarantee ethical research. *The Conversation*, 3 December https://theconversation.com/crispr-babies-raise-an-uncomfortable-reality-abiding-by-scientific-standards-doesnt-guarantee-ethical-research-108008.

Hurlbut, J. B. (2019). Human genome editing: ask whether, not how. *Nature*, 565, 135.

Hurlbut, J. B. (2021). Decoding the CRISPR-baby stories. *MIT Technology Review*, 24 February. www.technologyreview.com/2021/02/24/1017838/crispr-baby-gene-editing-jiankui-history/.

IBEF (2021). Indian Pharmaceuticals Industry Analysis. India Brand Equity Foundation, 22 March. www.ibef.org/industry/science-and-technology.aspx.

ICMR-DBT (2013). National Guidelines for Stem Cell Research. The Indian Council of Medical Research (ICMR).

India Today (2015). India becomes world's largest producer of cotton. *India Today*, 3 October. www.indiatoday.in/education-today/gk-current-affairs/story/largest-producer-of-cotton-266164-2015-10-03.

Indira, A., Bhagavan M., and Virgin, I. (2005). *Agricultural Biotechnology and Biosafety in India: Expectations, Outcomes and Lessons*. Stockholm Environment Institute, Stockholm.

Inter-Academy Report (2010). Inter-Academy Report on GM Crops. Indian Academy of Sciences, New Delhi. Indian National Science Academy (INSA). www.insaindia.res.in/pdf/Updated_Inter_Academy_Report_on_GM_crops.pdf.

IANR (International Association of Neurorestoratology) (2009). Beijing declaration of international association of neurorestoratology. *Neuroscience Bulletin*, 25(4): 228.

Irwin, A. (2018). No PhDs needed: how citizen science is transforming research. *Nature*, 562, 480–2.

Ivaldi G., Lanzone, E., and Woods, D. (2017). Varieties of populism across a left-right spectrum: the case of the Front National, the Northern League, Podemos and Five Star Movement. *Swiss Political Science Review*, 23(4), 354–76.

Japan Prime Minister's Office (2014). Japan Revitalization Strategy: Japan's Challenge for the Future. Kantei, 24 June. www.kantei.go.jp/jp/singi/keizaisaisei/pdf/honbunEN.pdf.

Jasanoff, S. (ed.) (2004). *States of Knowledge: The Co-production of Science and Social Order*. London: Routledge.

Jasanoff, S. (2005a). *Designs on Nature: Science and Democracy in Europe and the United States*. Princeton, NJ: Princeton University Press.

Jasanoff, S. (2005b). Judgment under siege: the three-body problem of expert legitimacy. In P. Weingart and M. Sabine (eds), *Democratization of Expertise? Exploring Novel Forms of Scientific Advice in Political Decision-Making*. Dordrecht: Kluwer.

Jasanoff, S. (2012). *Science and Public Reason*. London: Routledge.

Jasanoff, S. (2015). Future imperfect: science, technology, and the imaginations of modernity. In S. Jasanoff and S.-H. Kim (eds), *Dreamscapes of Modernity: Sociotechnical Imaginaries and the Fabrication of Power*. Chicago: University of Chicago Press.

Jasanoff, S., and Simmet, H. R. (2017). No funeral bells: public reason in a 'post-truth' age. *Social Studies of Science*, 47(5), 751–70.

Jasanoff, S., Hurlbut, J. B., and Saha, K. (2019). Democratic governance of human germline genome editing. *The CRISPR Journal*, 2, 266–71.

Jayaraman, K. S. (2003). Indian prime minister pledges to revamp science. *Nature*, 421, 101.

Jayaraman, K. S. (2005a). Biotech boom. *Nature*, 436(7050), 480–3.

Jayaraman, K. S. (2005b). Indian regulations fail to monitor growing stem-cell use in clinics. *Nature*, 434, 259.

Jayaraman, K. S. (2010). India's transgenic aubergine in a stew. Nature. 10 February 2010 https://doi.org/10.1038/news.2010.65.

Jayaraman, K. S. (2012). Indian science in need of overhaul, *Nature News*, 6 January. www.nature.com/news/indian-science-in-need-of-overhaul-1.9750.

Jeffery, R. (1979). Recognizing India's doctors: the institutionalization of medical dependency, 1918–39. *Modern Asian Studies*, 13(2), 301–26.

Jia, H. (2020). The seat of science capital. *Nature*, 585, S52–S54.

Jian, M. (2020). The vaccine war: from Ebola to Covid. *Zhishifenzi*, 26 May. http://zhishifenzi.com/depth/depth/9146.html.

Jin, B., Lin, L., and Rousseau, R. (2004). Long-term influences of interventions in the normal development of science: China and the cultural revolution. *Journal of the American Society for Information Science and Technology*, 55(6), 544–50.

Jishnu, L. (2011). Scientifically invalid. DowntoEarth.org, 31 January. www.downtoearth.org.in/news/scientifically-invalid-32901.

Joseph, M., and Robinson, A. (2014). Free Indian science. *Nature*, 508, 36–8.

Judson, H. F. (2006). The problematical Dr. Huang Hongyun. *MIT Technology Review*, 16 February. www.technologyreview.com/s/405327/the-problematical-dr-huang-hongyun/.

Kaiser, J., and Normal, D. (2015). Chinese paper on embryo engineering splits scientific community. *Science*. www.sciencemag.org/news/2015/04/chinese-paper-embryo-engineering-splits-scientific-community.

Kapil, S. (2021). Agri share in GDP hit 20% after 17 years: economic survey. *DowntoEarth.org*, 29 January. www.downtoearth.org.in/news/agriculture/agri-share-in-gdp-hit-20-after-17-years-economic-survey-75271.

Kapur, D. (2007). The economic impact of international migration from India. In U. Tambar (ed.), *Movement of Global Talent: The Impact of High Skill Labor Flows from India and China*. Princeton, NJ: Policy Research Institute for the Region, Princeton University.

Kapur, M. (2021). The inconsistencies in India's Covid-19 vaccine approval system. *Quartz India*, 26 February. https://qz.com/india/1977748/why-did-india-approve-covaxin-covishield-not-pfizer-sputnik-v/.

Kasperson, R. E., et al. (1988). The social amplification of risk: a conceptual framework. *Risk Analysis*, 8, 177–87.

Kauli, V. (2005). AIIMS pioneers stem cell injection. *Times of India*, 24 February. https://timesofindia.indiatimes.com/city/delhi/aiims-pioneers-stem-cell-injection/articleshow/1031528.cms.

Kaur, R. (2008). *Assessment of Genetic Damage in Workers Occupationally Exposed to Pesticides in Various Districts of Punjab*, Department of Human Biology, Punjabi University, Patiala.

Keating, P., and Cambrosio, A. (2012). *Cancer on Trial: Oncology as a New Style of Practice*. Chicago: University of Chicago Press.

Kennedy, D. (2002). Editorial: the importance of rice. *Science*, 296 (5565), 13.

Keim, B. (2003). Out of sight, out of mind: how Harvard University exploited rural Chinese villagers for their DNA. *Genewatch*, 16, 10–11.

Khan, S. (2012). Colonial medicine and elite nationalist responses in India: conformity and contradictions. In P. Bala (ed.), *Contesting Colonial Authority: Medicine and Indigenous Responses in Nineteenth- and Twentieth-Century India*. Lexington, KY: Lexington Books.

King, N. B. (2002). Security, disease, commerce: ideologies of postcolonial global health. *Social Studies of Science*, 32, 763–89.

Kingdon, J. W. (1984). *Agendas, Alternatives, and Public Policies*. London: Longman Publishing Group.

Kirksey, E. (2021). *The Mutant Project: Inside the Global Race to Genetically Modify Humans*. Bristol: Bristol University Press.

Knoppers, B. M., and Chadwick. R. (1994). The Human Genome Project: under an international ethical microscope. *Science*, 265, 2035–6.

Knoppers, B. M., and Chadwick, R. (2005). Human genetic research: emerging trends in ethics. *Nature Reviews Genetics* 6(1), 75–9.

Kochhar, R. (2019). Frauds in Indian scientific research. *The Tribune* (Indian), 3 July. www.tribuneindia.com/news/archive/comment/frauds-in-indian-scientific-research-789742.

Koleva, G. (2012). Stem cells and the lawsuit that may shape our medical future. Forbes, 10 February. www.forbes.com/sites/gerganakoleva/2012/02/10/stem-cells-and-the-lawsuit-that-may-shape-our-medical-future/.

Koley, M., Goveas, J. J., and Chakraborty, M. (2020). Open access: a problem way beyond one nation one subscription. *Times of India*, 13 December. https://timesofindia.indiatimes.com/blogs/voices/open-access-a-problem-way-beyond-one-nation-one-subscription/.

Kolte, D. B. (2020). Patenting of traditional knowledge and its relevance (intellectual property rights). *International Journal of Innovative Science and Research Technology*, 5, 178–80.

Krishna, V. V. (1991). Changing policy in science and tecnology in India. In R. Arvanitis (ed.), *Science & Technology Policy. Volume II*. Oxford: Eolss Publishers/UNESCO.

Krishna, V. V. (2001). Changing policy cultures, phases and trends in science and technology in India. *Policy Cultures*, 28(3), 179–94.

Krishna, V. V. (2013). Science, technology and innovation policy 2013: high on goals, low on commitment. *Economic and Political Weekly*, 48(16), 15–19.

Krishnan, A. (2021). China says no vaccine competition, but its media takes aim at India. *The Hindu*, 27 January. www.thehindu.com/news/international/china-says-no-vaccine-competition-but-its-media-takes-aim-at-india/article33678450.ece.

Kronstadt, K. A. (2004). CRS Report for Congress. India's 2004 National Elections. Library of Congress Washington DC Congressional Research Service, 7 December. https://apps.dtic.mil/sti/pdfs/ADA454697.pdf.

Kuhn, T. S. (1962 [1996]). *The Structure of Scientific Revolutions*. Chicago: University of Chicago Press.

Kumar, R. V. (2005). UK woos Indian investors. *The Hindu Businessline*. www.thehindubusinessline.com/todays-paper/tp-economy/uk-woos-indian-investors/article2180815.ece.

Kumar, R. (2016). *Rethinking Revolutions: Soyabean, Choupals and the Changing Countryside in Central India*. New Delhi: Oxford University Press.

Kumari, B., et al. (2004). Monitoring of pesticidal contamination of farmgate vegetables from Hisar. *Environmental Monitoring and Assessment*, 90(1), 65–71.

Laclau, E. (2005a). *On Populist Reason*. London: Verso.

Laclau, E. (2005b). Populism: what's in a name? In F. Panizza (ed.), *Populism and the Mirror of Democracy*. London: Verso.

Lal, S., and Malaviya, D. (2013). Biotechnology Regulatory Authority Bill 2013: an attempt to identify the obstacles to consensus on genetically modified crops. *Environment, Law and Society Journal*, 1, 137–46.

Lan, S., Yu, P., Wen, Q., and Zhang, L. (2020). 66 days of pandemic, 190 papers globally to decode the coronavirus. *National Business Daily*, 4 March. www.nbd.com.cn/articles/2020-03-04/1413612.html.

Lancet (2007). Animal–human hybrid-embryo research. www.thelancet.com/journals/lancet/article/PIIS0140673607614202/fulltext.

Landrain, T., Meyer, M., Perez, A. M., and Sussan, R. (2013). Do-it-yourself biology: challenges and promises for an open science and technology movement. *Systems and Synthetic Biology*, 7(3), 115–26.

Lasco, G. (2020). Medical populism and the COVID-19 pandemic. *Global Public Health*, 15, 1417–29.

Lasco, G., and Curato, N. (2019). Medical populism. *Social Science & Medicine*, 221(1), 1–8.

Lasco, G., and Larson, H. J. (2020). Medical populism and immunisation programmes: illustrative examples and consequences for public health. *Global Public Health*, 15, 334–44.

Latour, B. (1993). *We Have Never Been Modern*, trans. Catherine Porter. Cambridge, MA: Harvard University Press.

Laurance, J. (2011). Stem cells 'allow paralysed man to breathe again'. *Independent*, 23 October. www.independent.co.uk/life-style/health-and-families/health-news/stem-cells-allow-paralysed-man-to-breath-again-835222.html.

Lavakare, P. J. (2013). India today: from brain drain to brain gain. In I. Alon, V. Jones, and J. R. McIntyre (eds), *Business Education in Emerging Markets*. Basingstoke: Palgrave Macmillan.

Lawler, A. (2002). U.S. questions Harvard research in China. *Science*, 296(5565), 28.

Lawrence, C. (1985). Incommunicable knowledge: science, technology and the clinical art in Britain 1850–1914. *Journal of Contemporary History*, 20(4), 503–20.

Lee, C-S., and Schrank, A. (2010). Incubating innovation or cultivating corruption? The developmental state and the life sciences in Asia, *Social Forces*, 88(3), 1231–55.

Lei, R., Zhai, X., Zhu, W., and Qiu, R. (2019). Reboot ethics governance in China, *Nature*, 569 (7755), 184–6.

Lei, X. (2021). Chinese vaccine helping developing countries in the war against the pandemic. *Global Times*, 25 January. www.sohu.com/a/446542004_162522.

Leng, Z., et al. (2012). IANR V and 9th GCNN Conference with 4th ISCITT Symposium. *CNS & Neurological Disorders – Drug Targets*, 11(6), 643–6.

Leong, C. C., Jarvis, D., Howlett, M., and Migone, A. (2011). Controversial science-based technology public attitude formation and regulation in comparative perspective: the state construction of policy alternative in Asia. *Technology is Society*, 33, 128–36.

Levidow L., and Carr S. (1997). How biotechnology regulation sets a risk/ethics boundary. *Agriculture and Human Values*, 14, 29–43.

Li, E., Ji, C., and Wang, Y. (2011). Review the past, incentivise the present, prospecting the future. *Chinese Medical Ethics*, 24, 827–9.

Li, H., and Yan, Z. (2018). He Jiankui, creator of the global first gene-edited babies. *21st Century Business Review*, 27 November. www.21cbr.com/article/80135.html.

Li, N. (2013). Regulatory and ethical concerns evoked by the 'Golden Rice'. *Science & Technology Review* (Chinese), 30, 9.

Li, Y., and Guan, X. (2021). Rich countries buys 70% of global vaccine, China is providing vaccine aids to 53 developing countries. China News, 20 February. https://m.chinanews.com/wap/detail/zw/gn/2021/02-23/9416882.shtml.

Li, Y., Tabakow, P., Li, D., and Huang H. (2018). Commemorating Geoffrey Raisman: a great neuroscientist and one of the founders of neurorestoratology and the IANR. *Journal of Neurorestoratology*, 6, 29–39.

Liang, P., et al. (2015). CRISPR/Cas9-mediated gene editing in human tripronuclear zygotes. *Protein & Cell*, 6, 363–72.

Lin, W-Y., and Law, J. (2014). A correlative STS: lessons from a Chinese medical practice. *Social Studies of Science*, 44, 801–24.

Lindvall, O., and Hyun, I. (2009). Medical innovation versus stem cell tourism. *Science*, 324, 1664–5.

Linkova, M., and Stockelova, T. (2012). Public accountability and the politicisation of science: the peculiar journey of Czech research assessment. *Science and Public Policy*, 39, 618–29.

Listed Company Research Institute (2019). Biomedical industry gap widens between China and the world, policy promotes accelerated reform in industry. Jinrongjie, 18 June. http://stock.jrj.com.cn/2019/06/18175227720791.shtml.

Liu, T. (2016). The Wei ZeXi case and its impact on the 10-billion cell therapy industry. Sina, 5 June. http://finance.sina.com.cn/chanjing/cyxw/2016-05-06/doc-ifxryhhh1708034.shtml.

Liu, Y (2016). Misconceptions of GM: unpacking the 'public attitude towards GM technology survey'. *Science and Technology Daily* (Chinese), 17 May. http://scitech.people.com.cn/n1/2016/0517/c1007-28355424.html.

Liu, Y., and Feng, Y. (2004). The story of 1%: where did Chinese basic research funding go? *Newsweek* (Chinese), 13 April. www.ebiotrade.com/newsf/2004-4/L2004412231447.htm.

Lloyd-Roberts, S. (2012). Have India's poor become human guinea pigs? BBC News. www.bbc.com/news/magazine-20136654.

Lord Saatchi (2014). Lord Saatchi Bill: we must liberate doctors to innovate. *The Telegraph* (UK), 26 January. http://medicalinnovationbill.co.uk/lord-saatchi-bill-we-must-liberate-doctorsto-innovate/.

Lovell-Badge, R. (2008). The regulation of human embryo and stem-cell research in the United Kingdom. *Nature Reviews Molecular Cell Biology*, 9, 998–1003.

LSEA (Life Sciences Economy Alliance) (2020). The Second Session of the 1st Belt and Road Life Sciences Economy Alliance General Assembly was successfully held at CNGB, Shenzhen. Belt and Road Initiative Life Sciences Economy Alliance, 16 December. www.brlsea.org/article/24.

Lynch, M. (2013). Ontography: investigating the production of things, deflating ontology. *Social Studies of Science*, 43, 444–62.

Ma, C., et al. (2014). Monitoring progress towards the elimination of measles in China: an analysis of measles surveillance data. *Bulletin of the World Health Organization*, 92, 340–7.

Ma, D. (2019). Boundary repair: science and enterprise at the Chinese Academy of Sciences. *Social Studies of Science*, 49, 381–402.

Ma, E., and Lynch, M. (2014). Constructing the East-West boundary: the contested place of a modern imaging technology technology in South Korea's dual medical system. *Science, Technology & Human Values*, 39, 639–65.

Ma, R. (2020). Writing a new chapter in the dragon-elephant tango. *Xinhua*, 13 October. www.xinhuanet.com/comments/2019-10/13/c_1125099007.htm.

Maier, Charles S. (2000). Consigning the twentieth century to history: alternative narratives for the modern era. *American Historical Review*, 105, 807–31.

Malhotra, I. (2010). Swallowing the humiliation. *Indian Express*, 12 July. http://archive.indianexpress.com/news/swallowing-the-humiliation/645168/3.

Mallapaty, S. (2020). India pushes bold 'one nation, one subscription' journal-access plan. *Nature*, 586, 181–2.

Mallapaty, S (2021a). 'Pregnant' male rat study kindles bioethical debate in China. *Nature*, https://doi.org/10.1038/d41586-021-01885-0.

Mallapaty, S (2021b). China's five-year plan focuses on scientific self-reliance. *Nature*, 591, 353–4.

Mancini, F., Van Bruggen, A. H., Jiggins, J. L., Ambatipudi, A. C., and Murphy, H. (2005). Acute pesticide poisoning among female and male cotton growers in India. *International Journal of Occupational and Environmental Health*, 11(3), 221–32.

Mandavilli, A. (2005). India. *Nature*, 436(7050), 477.

Mandavilli, A. (2006). Profile: Hui Zhen Sheng. *Nature Medicine*, 12, 265.

Mani, S. (2013). The science and technology innovation policy of 2013. *Economic & Political Weekly*, 9 March.

Marchione, M. (2018). Chinese researcher claims first gene-edited babies. Associated Press, 26 November. https://apnews.com/articl e/4997bb7aa36c45449b488e19ac83e86d.

Marlow, I., Mangi, F., and Lindberg, K. S. (2020). China is struggling to get the world to trust its vaccines. Bloomberg, 29 December. www. bloomberg.com/news/features/2020-12-28/china-s-struggling-to-get- the-world-to-trust-its-covid-vaccines.

Martin, K. J. (2014). The world's not ready for this: globalizing selective technologies. *Science, Technology & Human Values*, 39, 432–55.

Masood, E. (2019). South Asia's pivot towards China. *Nature*, 569, 24–7.

McCann, Philip (2019). UK research and innovation: a place-based shift? July 2019. Centre for Science, Technology and Innovation Policy, University of Cambridge. www.ifm.eng.cam.ac.uk/uploads/Research/ CSTI/UKRI_Place/McCann_-_UK_Research_and_Innovation_-_A_Place- Based_Shift_vFinal.pdf.

McCarthy, N. (2019). The countries leading the world in scientific publications. Statista, 20 December www.statista.com/chart/20347/ science-and-engineering-articles-published/.

McLellan, T. (2020). Impact, theory of change, and the horizons of scientific practice. *Social Studies of Science*, 51(1), 100–20.

McMahon, D. S. (2014). The global industry for unproven stem cell inter- ventions and stem cell tourism. *Tissue Engineering and Regenerative Medicine*, 11, 1–9

Meagher, K. M., Allyse, M. A., Master, Z., and Sharp, R. R. (2020). Reexamining the ethics of human gremlin editing in the wake of scandal. *Mayo Clinic Proceedings*, 95(2), 330–8.

Médecins Sans Frontières (2021). Press Release: Countries obstructing COVID-19 patent waiver must allow negotiations to start. Médecins Sans Frontières , 9 March. www.msf.org/countries-obstructing-covid-19- patent-waiver-must-allow-negotiations.

Mehra, P. (2008). Scotland woos more Indian investment. *The Hindu Businessline*. www.thehindubusinessline.com/todays-paper/tp-economy/ scotland-woos-more-indian-investment/article1623355.ece.

Mehta, A. (2013). Food sovereignty, GMOs and the BRAI Bill. Green- peace. 22 August 2013. www.greenpeace.org/india/en/story/405/ food-sovereignty-gmos-and-the-brai-bill/.

Mehta, R., and Gopalakrishnan, B. N. (2021). Diaspora key to success of India's new science, technology. *Innovation Policy*, 17 January. www. americanbazaaronline.com/2021/01/17/diaspora-key-to-success-of-indias- new-science-technology-innovation-policy-443861/.

Merz, S. (2020). Global trials, local bodies: negotiating difference and sameness in Indian for-profit clinical trials. *Science, Technology & Human Values*, 46, 882–905.

Meyer, J. D., Bernal, J., Charum, J., Gaillard, J., Granes, J., et al (1997). Turning brain drain into brain gain: the Colombian experience of the diaspora option. *Science, Technology & Society*, 2(2), 285–315.

Mignolo, W. D. (2009). Epistemic disobedience, independent thought and decolonial freedom. *Theory, Culture & Society*, 26, 159–81.

Mignolo, W. D. (2011). *The Darker Side of Western Modernity: Global Futures, Decolonial Options*. Durham, NC: Duke University Press.

Mignolo, W. D., and Walsh, C. E. (2018). *On Decoloniality: Concepts, Analytics, Praxis*. Durham, NC: Duke University Press.

Miller, H. L. (1996). *Science and Dissent in Post-Mao China: The Politics of Knowledge*. Seattle, WA: University of Washington Press.

Ministry of Agriculture, China (2010). Application prospect of phytase transgenic maize. Ministry of Agriculture, China, 17 July. www.moa.gov.cn/ztzl/zjyqwgz/kpxc/201007/t20100717_1601264.htm.

Ministry of Agriculture and Farmers Welfare, India (2019). Agriculture Census 2015–16. Ministry of Agriculture and Farmers Welfare, India. http://agcensus.nic.in/document/agcen1516/T1_ac_2015_16.pdf.

Ministry of Education China (2020). Statistics on 2019 Chinese Studying Overseas. Ministry of Education, 14 December. www.moe.gov.cn/jyb_xwfb/gzdt_gzdt/s5987/202012/t20201214_505447.html.

Ministry of Human Resources and Social Security, China (2017). Guiding Opinions of the Ministry of Human Resources and Social Security on Supporting and Encouraging Professional Technical Personnel of Public Institutions in Innovation and Entrepreneurship. Beijing: Ministry of Human Resources and Social Security, 10 March. www.mohrss.gov.cn/rydwrsgls/SYDWRSGLSzhengcewenjian/201703/t20170318_268143.html.

Ministry of Human Resources and Social Security, China (2019). Further Guiding Opinions of the Ministry of Human Resources and Social Security on Supporting and Encouraging Professional Technical Personnel of Public Institutions in Innovation and Entrepreneurship. *Beijing: Ministry of Human Resources and Social Security*, 27 December. www.mohrss.gov.cn/wap/zc/zcwj/202001/t20200120_356477.html.

Ministry of Science and Technology, China (2005). The Course of China's Science and Technology Development. Ministry of Science and Technology, 23 September. www.gov.cn/test/2005-09/23/content_69616.htm.

Ministry of Science and Technology, India (2013). Science, Technology and Innovation Policy. Ministry of Science and Technology. https://indiabioscience.org/media/articles/STIP-2013.pdf.

Mirowski, P. (2002). *Machine Dreams Economics Becomes a Cyborg Science.* Cambridge: Cambridge University Press.

Mirowski, P., and Sent, E. M. (2002). Introduction. In P. Mirowski and E. M. Sent (eds), *Science Bought and Sold: Essays in the Economics of Science.* Chicago: University of Chicago Press.

Mirowski, P., and van Horn, R. (2005). The contract research organization and the commercialisation of scientific research. *Social Studies of Science*, 35, 503–48.

Mishra, P. (2005). How India reconciles Hindu values and biotech. *New York Times*, 21 August. www.nytimes.com/2005/08/21/weekinreview/how-india-reconciles-hindu-values-and-biotech.html.

Mitchell, S. (2009). *Unsimple Truths: Science, Complexity, and Policy.* Chicago: University of Chicago Press.

Mitra, A. K. (2021). India's Bharat biotech pursues COVID-19 vaccine approval in over 40 countries. Reuters, 18 February. www.reuters.com/article/us-health-coronavirus-india-vaccine-idUSKBN2AI0MW.

MoEF-GEAC (2009). Genetic Engineering Approval Committee. 2009 Report of the Expert Committee (EC-II) on Bt Brinjal Event EE-1 Developed by: M/s Maharashtra Hybrid Seeds Comany Ltd. (MAHYCO), Mumbai; University of Agricultural Sciences (UAS), Dharwad; and Tamil Nadu Agriculture. Ministry of Environment and Forests (MoEF).

Moffitt, B. (2017). Transnational populism? Representative claims, media and the difficulty of constructing a transnational 'people'. *Javnost – The Public*, 24(4), 409–25.

MOH (Ministry of Health, China) (2000). Notice on MOH setting up medical ethics expert committee. *Ministry of Health*, 6 March. www.nhc.gov.cn/zwgkzt/pkjjy1/200805/35740.shtml.

MOH (Ministry of Health, China) (2007). Measures for the Ethical Review of Biomedical Research Involving Humans (Trial Version). Ministry of Health, January. www.chinacdc.cn/jdydc/200701/t20070129_32217.htm.

MOH (Ministry of Health, China) (2010). Press Conference on Parents' Refusal to Vaccinate their Children. Ministry of Health, 13 April. http://news.sohu.com/20100413/n271484230.shtml.

MOH (Ministry of Health, China) and MOST (Ministry of Science and Technology, China) (2003). Ethical Guidelines for Research on Human Embryonic Stem Cells. Beijing: MOH and MOST, 24 December.

Mokoena, R. (2007). South–South co-operation: the case for IBSA. *South African Journal of International Affairs*, 14(2), 125–45.

MOST (Ministry of Science and Technology, China) (1998). Interim Measures for the Management of Human Genetic Resources. Ministry of Science and Technology. www.most.gov.cn/fggw/xzfg/200811/t20081106_64877.htm.

MOST (Ministry of Science and Technology, China) (2002). Law of Science Popularisation. Ministry of Science and Technology, July. www.npc.gov.cn/wxzl/wxzl/2002–07/10/content_297301.htm.

MOST (Ministry of Science and Technology, China) (2005) *The Course of China's Science and Technology Development.* The Central People's Government of the People's Republic of China website. 23 September 2005. www.gov.cn/test/2005-09/23/content_69616.htm.

MOST (Ministry of Science and Technology, China) and MOF (Ministry of Finance, China) (2020). Science and Technology Transfer Year Book 2019 (Higher Education and Research Institute Report). Beijing: MOST and MOF.

Mudde, C. (2007). *Populist Radical Right Parties in Europe.* Cambridge: Cambridge University Press.

Mudde, C. (2016). *On Extremism and Democracy in Europe.* London: Routledge.

Mudde, C., and Kaltwasser, C. R. (2017). *Populism: A Very Short Introduction.* Oxford: Oxford University Press.

Mulkay, M. (1994). Changing minds about embryo research. *Public Understanding of Science,* 3(2), 195–213.

Müller, Ja-W. (2016). *What Is Populism?* Philadelphia, PA: University of Pennsylvania Press.

National Academies of Sciences, Engineering, and Medicine, US (2015). International Summit on Human Gene Editing: A Global Discussion. Washington, DC: The National Academies Press.

National Development and Reform Commission (2008). Long Term Plan for National Food Security (2008–20). Beijing: National Development and Reform Commission, 11 November. www.gov.cn/test/2008-11/14/content_1148698.htm.

National Health and Family Planning Commission of the People's Republic of China (2016). Measures for the Ethical Review of Biomedical Research Involving Humans. National Health and Family Planning Commission of the People's Republic of China. www.gov.cn/gongbao/content/2017/content_5227817.htm.

National Science Foundation, US (2018). Publication Output, by Region, Country, or Economy. National Science Foundation, US. https://ncses.nsf.gov/pubs/nsb20206/publication-output-by-region-country-or-economy.

Nature Editorial (2010). Do scientists really need a PhD? *Nature,* 464, 7.

Nature Index (2019). Top 100 institutions with high affiliation articles in genetics. Nature Index. www.natureindex.com/supplements/nature-index-2019-collaboration-and-big-science/tables/top100-genetics.

Nature Index (2021). Top 100 Rising Institutuions. Nature Index. www.natureindex.com/annual-tables/2020/institution/rising.

Nature Medicine (2006). People to watch: Hongyun Huang. *Nature Medicine*, 12, 262.

Nayar, B. R. (2005). India in 2004: regime change in a divided democracy. *Asian Survey*, 45(1), 71–82.

Needham, J. (1969). *The Grand Titration: Science and Society in East and West*. London: Allen & Unwin.

New York Times (1946). The tragic paradox of our age. *New York Times Magazine*, 7 September.

Niño-Zarazúa, M., and Addison, T. (2012). Redefining poverty in China and India. *United Nations University*. https://unu.edu/publications/articles/redefining-poverty-in-china-and-india.html.

Normile, D. (2020). China again boosts R&D spending by more than 10%. *Science*, 28 August.

Normile, D. (2018). CRISPR bombshell: Chinese researcher claims to have created gene-edited twins. *Science*, 26 November.

Novella, S. (2010). Cracking down on stem cell tourism. https://sciencebasedmedicine.org/cracking-down-on-stem-cell-tourism/.

Nowotny, H. (2007). How many policy rooms are there? Evidence-based and other kinds of science policies. *Science, Technology & Human Values*, 32, 479–90.

NSF (National Science Foundation) (2020) Publication Output, by Region, Country, or Economy. https://ncses.nsf.gov/pubs/nsb20206/publication-output-by-region-country-or-economy.

Nundy, S., and Gulhati, C. M. (2005). A new colonialism? Conducting clinical trials in India. *The New England Journal of Medicine*, 352(16), 1633–6.

Office of the Leading Group for Promoting the BRI (2019). The Belt and Road Initiative: Progress, Contributions and Prospects. Belt and Road Portal, 22 April. https://eng.yidaiyilu.gov.cn/zchj/qwfb/86739.htm.

Olesen, A. (2015). China has its own anti-vaxxers: blame the internet. *Foreign Policy*, 16 March. https://foreignpolicy.com/2015/03/16/china-has-its-own-anti-vaxxers-blame-the-internet/.

Oliver, A. L. (2004). Biotechnology entrepreneurial scientists and their collaborations. *Research Policy*, 33(4), 583–97.

Ong, A. (2006). *Neoliberalism as Exception: Mutations in Citizenship and Sovereignty*. Durham, NC: Duke University Press.

Ong, A. (2012). Powers of sovereignty: state, people, wealth, life. *Focaal*, 64, 24–35.

Ong, A. (2016) *Fungible Life: Experiment in the Asian City of Life*. Durham, NC: Duke University Press.

Ong, A., and Chen, N. (eds) (2010). *Asian Biotech: Ethics and Communities of Fate*. Durham, NC: Duke University Press.

Ouyang, G. (2003). Scientism, technocracy, and morality in China. *Journal of Chinese Philosophy*, 30, 177–93.

Overseas Intel (2020). Chinese genius Cao Yuan: American Green Card and a Chinese destiny to returns. *Sohu*, 30 April. www.sohu.com/a/392298008_194632.

Padma, T. V. (2015). India: the fight to become a science superpower. *Nature*, 521, 144–7.

Page, J., and Hinshaw, D. (2021). China refuses to give WHO raw data on early Covid-19 cases. *Wall Street Journal*, 12 February. www.wsj.com/articles/china-refuses-to-give-who-raw-data-on-early-covid-19-cases-11613150580.

Palladino, P., and Worboys, M. (1993). Science and imperialism. *Isis*, 84, 91–102.

Pandey, A., and Sharma, M. (2021). Bharat biotech top boss slams criticism against indigenous vaccine, defends emergency authorization. *India Today*, 4 January. www.indiatoday.in/coronavirus-outbreak/vaccine-updates/story/bharat-biotech-cmd-covaxin-emergency-approval-1755783-2021-01-04.

Pandya, S. (2008). Stem cell transplantation in India: tall claims, questionable ethics. *Indian Journal of Medical Ethics*, 5(1), 15–18.

Parliament of India (2005a). Rajya Sabha proceedings at The Parliament House. Rajya Sabha (Upper House), 5 May. http://rajyasabha.gov.in/rsnew/session_journals/204/06052005.pdf.

Parliament of India (2005b). Department-Related Parliamentary Standing Committee on Health and Family Welfare. Fifteenth Report on Action Taken by Government on the Recommendations/Observations Contained in the Seventh Report on Demands For Grants 2005–2006 (Demands No. 47). Parliament of India.

Pasricha, A. (2021). India launches 'neighborly vaccine diplomacy'. *Voice of America*, 24 January. www.voanews.com/covid-19-pandemic/india-launches-neighborly-vaccine-diplomacy.

Pawar, Y. (2012). Govt bans Mahyco Monsanto Biotech. *DNA India*. www.dnaindia.com/mumbai/report-govt-bans-mahyco-monsanto-biotech-1726118.

PCI (2002). 10th Five Year Plan. As Approved by CABINET. Government of India, Planning Commission, Yojana Bhavan, New Delhi, 1101–3. http://planningcommission.gov.in/plans/planrel/fiveyr/10th/volume2/v2_ch10_1.pdf.

PCI (2007). 11th Five Year Plan. As Approved by CABINET. Government of India, Planning Commission, Yojana Bhavan, New Delhi, 176–9. http://planningcommission.gov.in/plans/planrel/fiveyr/11th/11_v1/11v1_ch8.pdf.

PCI (2011). 12th Five Year Plan. As Approved by CABINET. Government of India, Planning Commission, Yojana Bhavan, New Delhi, 244–53. http://planningcommission.gov.in/plans/planrel/fiveyr/12th/pdf/12fyp_vol1.pdf.

Peel, M., Fleming, S., Parker, G., and Foster, P. (2021). Brussels hits back in vaccine exports row with UK. *Financial Times*, 10 March. www.ft.com/content/d8084bca-2310-4994-b034-87a4e2fc386d.

Petersen, A., Tanner C., and Munsie, M. (2015). Between hope and evidence: how community advisors demarcate the boundary between legitimate and illegitimate stem cell treatments. *Health*, 19, 188–206.

Petryna, A. (2005). Ethical variability: drug development and globalizing clinical trials. *American Ethnologist*, 32, 183–97.

Petryna, A. (2009). *When Experiments Travel: Clinical Trials and the Global Search for Human Subjects*. Princeton, NJ: Princeton University Press.

Plagemann, J., and Destradi, S. (2019). Populismus and foreign policy: the case of India. *Foreign Policy Analysis*, 15, 283–301.

Powell, W. (2004). Knowledge networks as a channels and conduits: the effects of spillovers in the Boston biotechnology community. *Organization Science*, 15, 5–21.

Pradhan, B., and Sen, S. R. (2021). India has plenty of coronavirus vaccines but few takers. Bloomberg, 26 January. www.bloomberg.com/news/articles/2021-01-26/india-s-unusual-vaccine-problem-plenty-of-shots-but-few-takers.

Prakash, G. (1999). *The Sign of Science. In Another Reason: Science and the Imagination of Modern India*. Princeton, NJ: Princeton University Press.

Prasad, A. (2005). Scientific culture in the 'other' theater of modern science: an analysis of the culture of magnetic resonance imaging research in India. *Social Studies of Science*, 35, 463–89.

Prasad, A. (2017). Biopolitical excess: techno-legal assemblage of stem cell research in India. *Science, Technology & Society*, 22, 102–23.

Prasad, C. S. (2008). Knowledge, democracy and science policy: the missing dialogue in globalised India. *IUP Journal of Governance and Public Policy*, 3(2–3), 87–102.

Prasad, C. S. (2020). Constructing alternative socio-technical worlds: re-imagining RRI through SRI in India. *Science, Technology & Society*, 25, 291–307.

Prasad, K. V. (2018). China-India Plus can be new model in South Asia. *The Tribune* (Indian), 30 September. www.tribuneindia.com/news/archive/nation/-china-india-plus-can-be-new-model-in-south-asia-661149.

Pratt, M. L. (1992). *Imperial Eyes: Travel Writing and Transculturation*. London: Routledge.

PTI (2005). Clinic's embryonic stem cell therapy worries govt. Press Trust of India. www.rediff.com/news/2005/nov/16stem.htm.

PTI (2021). India to play vital role in equitable distribution of COVID-19 vaccines around the world. *The Hindu*, 27 December. www.thehindu.com/business/Industry/india-to-play-vital-role-in-equitable-distribution-of-covid-19-vaccines-around-the-world-pharma-industry/article33430054.ece.

Qiu, J. (2007). To walk again. *New Scientist*, 10 November 2007, 57–9.

Qiu, J. (2009). China spinal cord injury network: changes from within. *Lancet*, 8, 606–7.

Qiu, J. (2012). China sacks officials over Golden Rice controversy. *Nature News*, 10.

Qiu, J. (2020). How China's 'Bat Woman' hunted down viruses from SARS to the new coronavirus. *Scientific American*, June issue. www.scientificamerican.com/article/how-chinas-bat-woman-hunted-down-viruses-from-sars-to-the-new-coronavirus1/.

Qiu, R. (2012). The path of bioethics in China: in commemoration of bioethics development in China. *Chinese Medical Ethics*, 25, 3–6.

Qiu, R., and Zhai, X. (2013). On the ethics and regulatory issue of ethics review committee. *Chinese Medical Ethics*, 26, 545–50.

Quijano, A. (2000). Coloniality of power, Eurocentrism, and Latin America. *Nepantla: Views from the South*, 1(3), 533–80.

Raina, R. (2013). Knowing and administering food: how do we explain persistence? Keynote talk at the National Seminar on Food Security and Food Production Institutional Challenges in the Governance Domain, organized by the Indian Institute of Public Administration (IIPA), New Delhi 31 October–1 November 2013.

Rajan, K. S. (2006). *Biocapital: The Constitution of Postgenomic Life*. Durham, NC: Duke University Press.

Rajão, R., Duque, R. B., and De', R. (2014). Introduction: voices from within and outside the South – defying STS epistemologies, boundaries, and theories. *Science, Technology & Human Values*, 39, 767–72.

Ramasubban, R. (2007). History of public health in modern India: 1857–2005. In M. L. Lewis and K. L. MacPherson (eds), *Public Health in Asia and the Pacific: Historical and Comparative Perspectives*. London: Routledge.

Ramesh, R. (2005). Row over doctor's 'miracle cures'. *Guardian*, 18 November. www.theguardian.com/science/2005/nov/18/stemcells.controversiesinscience.

Rankin, W. (2017). Zombie projects, negative networks, and multigenerational science: the temporality of the International Map of the World. *Social Studies of Science*, 47, 353–75.

Rao, Y. (2012). Why Huanming Yang is a gangster entrepreneur. *EBiotrade*, November. www.ebiotrade.com/newsf/2012-11/201211293810406.htm.

Ratchford, J. T., and Blanpied, W. A. (2008). Path to the future for science and technology in China, India and the United States. *Technology in Society*, 30, 211–33.

Rawal, N., and Bai, Z. (2021). Could India and China help stop 'vaccine nationalism'? *World Economic Forum*, 9 March. www.weforum.org/agenda/2021/03/india-china-help-covax-fulfill-vaccine-mission/.

Regal, B. (2018). Demythologising the 'liberalisation' of clinical trial regulations in India: a practitioner's perspective. *Science, Technology & Society*, 23, 463–84.

Regalado, A. (2018). Chinese CRISPR scientist cited US report as his green light. *MIT Technology Review*, 27 November. www.technologyreview.com/2018/11/27/1821/rogue-chinese-crispr-scientist-cited-us-report-as-his-green-light/.

Regalado, A. (2019). The DIY designer baby project funded with Bitcoin. *MIT Technology Review*, 1 February. www.technologyreview.com/2019/02/01/239624/the-transhumanist-diy-designer-baby-funded-with-bitcoin/.

Ren, J., et al. (2014). Similar challenges but different responses: media coverage of measles vaccination in the UK and China. *Public Understanding of Science*, 23(4), 366–75.

Ren, Y., and Wan, P. (eds) (2020). Understanding Xi Jinping's three key points on science, innovation and development. *The Communist Party of China News*, 18 September. http://theory.people.com.cn/n1/2020/0918/c40531-31866102.html.

Reuters (2021). What's behind varying efficacy data for Sinovac's COVID-19 vaccine? *Reuters*, 14 January. www.reuters.com/article/health-coronavirus-sinovac-explainer-int-idUSKBN29J0M1.

Rickard, L. N. (2011). In backyards, on front lawns: examining informal risk communication and communicators. *Public Understanding of Science*, 20(5), 642–57.

RIS (Research and Information System for Developing Countries) (2016). Trinity for Development, Democracy and Sustainability. New Delhi: Research and Information System for Developing Countries.

Robels, R. (2021). Scepticism over China's Sinovac jab as Philippines rolls out coronavirus vaccination programme. *South China Morning Post*, 1 March. www.scmp.com/week-asia/politics/article/3123643/scepticism-over-chinas-sinovac-jab-philippines-rolls-out.

Rodrik, D. (2018). Populism and the economics of globalization. *Journal of International Business Policy*, 1–22.

Rogaski, R. (2014). *Hygienic Modernity: Meanings of Health and Disease in Treaty-Port China*. Berkeley, CA: University of California Press.

Roosevelt, M. (2000). Inside the protests: taking it to Main Street. *TIME*, 31 July. http://content.time.com/time/subscriber/printout/0,8816,997621,00.html#.

Rosemann, A. (2013). Scientific multi polarisation: its impact on international clinical research collaborations and theoretical implications. *Science, Technology & Society*, 18, 339–59.

Rosemann, A., and Chaisinthop, N. (2016). The pluralization of the international: resistance and alter-standardisation in regenerative stem cell medicine. *Social Studies of Science*, 46, 112–39.

Rosemann, A., and Sleeboom-Faulkner, M. (2016). New regulation for clinical stem cell research in China: expected impact and challenges for implementation. *Regenerative Medicine*, 11, 5–9.

Roudometof, V. (2019). Recovering the local: from globalisation to localization. *Current Sociology Review*, 67, 801–17.

Roy, A., Rocha, E., and Das, K. N. (2020). Not without India: world's pharmacy gears up for vaccine race. Reuters, 10 December. www.reuters.com/article/health-coronavirus-india-vaccine/not-without-india-worlds-pharmacy-gears-up-for-vaccine-race-idINKBN28K10E?edition-redirect=uk.

Russo, E. (2005). Follow the money: the politics of embryonic stem cell research. *PLoS Biology*, 3(7), e234.

Sabharwal, M., and Varma, R. (2017). Convergence or divergence: practice of science by migrant faculty in India and the United States. *Science, Technology & Human Values*, 42, 775–94.

Safi, M. (2021). Most poor nations will take until 2024 to achieve mass Covid-19 immunisation. *Guardian*, 27 January. www.theguardian.com/society/2021/jan/27/most-poor-nations-will-take-until-2024-to-achieve-mass-covid-19-immunisation.

Said, E. (1978). *Orientalism*. New York: Vintage.

Salter, B. (2008). Governing stem cell science in China and India: emerging economies and the global politics of innovation. *New Genetics and Society*, 27(2), 145–59.

Salter B., and Jones, M. (2002). Human genetic technologies, European governance and the politics of bioethics. *Nature Reviews Genetics*, 3, 808–14.

Salter, B., and Jones, M. (2005). Biobanks and bioethics: the politics of legitimation. *Journal of European Public Policy*, 12(4), 710–32.

Salter, B., and Salter, C. (2007). Bioethics and the global moral economy: the cultural politics of human embryonic stem cell science. *Science, Technology & Human Values*, 32, 554–81.

Salter, B., Zhou, Y., Datta, S., and Salter, C. (2016). Bioinformatics and the politics of innovation in the life sciences: science and the state in the United Kingdom, China and India. Science. *Technology & Human Values*, 41, 793–826.

Sangwan, V. S., et al. (2005). Early results of penetrating keratoplasty following limbal stem cell transplantation. *Indian Journal of Ophthalmology*, 53, 31–5.

Santos, B. de Sousa. (2002). *Toward a New Legal Common Sense: Law, Globalization, and Emancipation*. London: Butterworths.

Santos, B. de Sousa (2014). *Epistemologies of the South: Justice Against Epistemicide*. Boulder, CO: Paradigm.

Santos, B. de Sousa, and Meneses, M. P. (eds) (2020). *Knowledge Born in the Struggle: Constructing the Epistemologies of the Global South*. New York: Routledge.

Sariola, S., and Simpson, B. (2019). *Research as Development: Biomedical Research, Ethics and Collaboration in Sri Lanka*. Ithaca, NY: Cornell University Press.

Sariola, S., Deapica, R., Anand K., and Jeffery, R. (2015). Big-pharmaceuticalisation: clinical trials and contract research organizations in India. *Social Science & Medicine* 131 (C), 239–46.

Saran, S. (2019). China's emergence as a scientific power. *Business Standard*, 7 May, https://cprindia.org/sites/default/files/China%27s%20emergence%20as%20a%20scientific%20power.pdf.

Sassen, S. (1991). *The Global City: New York, London, Tokyo*. Princeton, NJ: Princeton University Press.

Sassen, S. (2006). *Territory, Authority, Rights: From Medieval to Global Assemblages*. Princeton, NJ: Princeton University Press.

Schurman, R., and Munro, W. A. (2013). *Fighting for the Future of Food: Activists versus Agribusiness in the Struggle over Biotechnology*. Minneapolis, MN: University of Minnesota Press.

ScienceNet (2018). Joint statement from 122 scientists: strong condemnation on the first HIV gene editing. ScienceNet.cn, 26 November. http://news.sciencenet.cn/htmlnews/2018/11/420386.shtm.

Scoones, I. (2006). *Science, Agriculture and the Politics of Policy: The Case of Biotechnology in India*. Delhi: Orient Blackswan.

Seah, S., et al. (2021). *The State of Southeast Asia: 2021*. Singapore: ISEAS-Yusof Ishak Institute.

Segal, A. (2011). *Advantage: How American Innovation Can Overcome the Asian Challenge*. New York: W. W. Norton.

Sevastopulo, D., and Hornby, L. (2014). Chinese environmental protest broken up. *Financial Times*, 31 March. www.ft.com/content/aa592b90-b89d-11e3-a189-00144feabdc0.

Shanbhag, S. S., Patel, C., Goyal, R., Donthineini, P. R., Singh, V., and Basu, S. (2019). Simple limbal epithelial transplantation (SLET): review of indications, surgical technique, mechanism, outcomes, limitations, and impact. *Indian Journal of Ophthalmology*, 67, 1265–77.

Shapin, S., and Schaffer, S. (1985). *Leviathan and the Air Pump*. Princeton, NJ: Princeton University Press.

Sharma, A. (2006). Stem cell research in India: emerging scenario and policy concerns. *Asian Biotechnology and Development Review*, 8(3), 43–53.

Sharma, D. (1976). Growth and failures of India's science policy. *Economic and Political Weekly*, 11(51), 1969–71.

Sharma, S. (2011). India in 2010: robust economics amid political stasis. *Asian Survey*, 51(1), 111–24.

Sharma, H. S., and Sharma, A. (2011). 8th Annual Conference of the Global College of Neuroprotection and Neuroregeneration. *Expert Review of Neurotherapeutics*, 11, 1121–4.

Sharp, J. P. (2009). *Geographies of Postcolonialism: Spaces of Power and Representation*. London: Sage.

Shen, S. X. (2016). Negotiating authorship in Chinese universities: how organisations shape cycles of credit in science. *Science, Technology & Human Values*, 41, 660–85.

Shen, X., and Williams, R. (2005). A critique of China's utilitarian view of science and technology. *Science, Technology & Society*, 197–214.

Shepard, W. (2017). How India got wrapped up in China's Belt and Road Initiative, despite opposing it. Forbes, 25 January. www.forbes.com/sites/wadeshepard/2017/06/25/china-now-has-india-wrapped-up-in-its-belt-and-road/?sh=773268af227f.

Shiva, V. (1997). *Biopiracy: The Plunder of Nature and Knowledge*. Cambridge MA: South End Press.

Shou, P., and Li, H. (2001). Huge dispute over China-US collaboration, heated gene ethics debate at Hangzhou conference. *Southern Weekly*, 6 April. www.china.com.cn/txt/2001-04/06/content_5027995.htm.

Shroff, G. (2018). Establishment and use of injectable human embryonic stem cells for clinical application. In A. Bharadwaj (ed.), *Global Perspective on Stem Cell Technologies*. Basingstoke: Palgrave.

Shroff, G., and Barthakur, J. K. (2015). Safety of human embryonic stem cells in patients with terminal/incurable conditions – a retrospective analysis. *Annals of Neurosciences*, 22, 132–8.

Shrum, W., et al. (2020). Who's afraid of Ebola? Epidemic fires and locative fears in the Information Age. *Social Studies of Science*, 50, 707–27.

Simmel, G. (1950). The stranger. In K. Wolff (ed.), *The Sociology of Georg Simmel*. New York: Free Press of Glencoe.

Simpson, B. (2018). A 'we' problem for bioethics and the social sciences: a response to Barbara Prainsack. *Science, Technology & Human Values*, 43, 45–55.

Singh, Z. D. (2019). Rethinking India's approach to China's Belt and Road Initiative, *Economic & Political Weekly*, 29 June.

Sinhal, K. (2008). 49 babies die during clinical trials at AIIMS. *Times of India*, 18 August. https://timesofindia.indiatimes.com/india/49-babies-die-during-clinical-trials-at-AIIMS/articleshow/3374492.cms.

Sipp, D. (2009). The rocky road to regulation. *Nature Reports Stem Cells*. http://doi.org/10.1038/stemcells.2009.125.

SKY News. (2006). Stem Cell Miracles. 23 January.

SKY News. (2007). Miracle Stem Cell Cure? 13 April.

Slater, D. (1997). *Consumer Culture and Modernity*. Cambridge: Polity.

Sleeboom-Faulkner, M. (2019). Regulatory brokerage: competitive advantage and regulation in the field of regenerative medicine. *Social Studies of Science*, 49, 355–80.

Sleeboom-Faulkner, M., and Hwang, S. (2012). Governance of stem cell research: public participation and decision-making in China, Japan, South Korea and Taiwan. *Social Studies of Science*, 42, 684–708.

Sleeboom-Faulkner, M., and Patra, P. K. (2011). Experimental stem cell therapy: biohierarchies and bionetworking in Japan and India. *Social Studies of Science*, 44(41), 645–66.

Smallman, M. (2018). Science to the rescue or contingent progress? Comparing 10 years of public, expert and policy discourses on new and emerging science and technology in the United Kingdom. *Public Understanding of Science*, 27(6), 655–73.

Smallman, M. (2020). 'Nothing to do with the science': How an elite sociotechnical imaginary cements policy resistance to public perspectives on science and technology through the machinery of government. *Social Studies of Science*, 50, 589–608.

Song, P. (2017). *Biomedical Odysseys: Fetal Cell Experiments from Cyberspace to China*. Princeton, NJ: Princeton University Press.

Sood, J. (2013). Monsanto told to quit India. DowntoEarth.org, 8 August. www.downtoearth.org.in/news/monsanto-told-to-quit-india-41878.

Sood, J. (2012). Whose germplasm is it? DowntoEarth.org, 15 October. www.downtoearth.org.in/coverage/whose-germplasm-is-it-39206.

Spivak, G. C. (1988). Can the subaltern speak? In C. Nelson and L. Grossberg (eds), *Marxism and the Interpretation of Culture*. Basingstoke: Palgrave Macmillan.

Spivak, G. C. (1999). *A Critique of Postcolonial Reasons: Towards a History of the Vanishing Present*. Cambridge. MA: Harvard University Press.

State Council, China (1995). General Planning for the 211 Project. Beijing: State Council.

State Council, China (1996). Decision of the State Council Concerning the Deepening of the Reform of the Science and Technology Management System. Beijing: State Council.

State Council, China (1999). Action Plan on Education Rejuvenation for the 21st Century. Beijing: State Council.

State Council, China (2006). Action Programme on National Science Literacy (2006–2010–2020). Beijing: State Council, 6 February www.gov.cn/gongbao/content/2006/content_244978.htm.

State Council, China (2011). Guiding Suggestions on Differentiated Public Institution Reform. Beijing: State Council, 23 March. www.gov.cn/gongbao/content/2012/content_2121699.htm.

State Council, China (2015). Opinions of the State Council on Several Policies and Measures for Vigorously Advancing the Popular Entrepreneurship and Innovation. *State Council*, 11 June. www.gov.cn/zhengce/content/2015-06/16/content_9855.htm.

Stavrakakis, Y., Katsambekis, G., Nikisianis, N., Kioupkiolis, A., and Siomos, T. (2017). Extreme right-wing populism in Europe: revisiting a reified association. *Critical Discourse Studies*, 14(4), 420–39.

Stone, G. D. (2007). Agricultural deskilling and the spread of genetically modified cotton in Warangal. *Current Anthropology*, 48, 67–103.

Stone, G., and Glover, D. (2011). Genetically modified crops and the 'food crisis': discourse and material impacts. *Development in Practice*, 21, 509–16.

Strathern, M. (1992). *Reproducing the Future: Essays on Anthropology, Kinship and the New Reproductive Technologies*. New York: Routledge.

Studer, S. (2021). Young Chinese are both patriotic and socially progressive. *The Economist*, 23 January. www.economist.com/special-report/2021/01/21/young-chinese-are-both-patriotic-and-socially-progressive.

Stuenkel, O. (2014). *India-Brazil-South Africa Dialogue Forum (IBSA): The Rise of the Global South*. Abingdon: Routledge.

Subramanian, A. K. (1981). Health for All: An Alternative Strategy: A Note on the Current Task A. IIMA Working Papers WP1981-11-01_00472, Indian Institute of Management Ahmedabad, Research and Publication Department.

Sun, Z. (2020). China establishes national science and technology ethics committee. *China News*, 21 October. www.chinanews.com/gn/2020/10-21/9319022.shtml.

Suo, Q. (2016). Chinese academic assessment and incentive system. *Science and Engineering Ethics*, 22(1), 297–9.

Suttmeier, R. P. (1974). *Research and Revolution: Science Policy and Societal Change in China*. Lexington, KY: Lexington Books.

Sutz, J. (2003). Inequality and university research agendas in Latin America. *Science, Technology & Human Values*, 28, 52–68.

Tang, Li., Hu, G., Sui, Y., Yang Y., and Cao, C. (2020). Retraction: the 'other face' of research collaboration? *Science and Engineering Ethics*, 26, 1681–708.

Tatlow, D. K. (2015). A scientific ethical divide between China and West. *New York Times*, 29 June. www.nytimes.com/2015/06/30/science/a-scientific-ethical-divide-between-china-and-west.html.

Teng, Y., and Feng, Z. (2013). Procedural justice in the right protection of human subjects: reflections on the Golden Rice controversy. *Medicine and Philosophy* (China), 34, 35–58.

Thorsteinsdottir, H., Ray, M., Kapoor, A., and Daar, A. S. (2011). Health biotechnology innovation on a global stage. *Nature Reviews: Microbiology*, 9, 137–43.

Tian, H., and Zhai, X. (2013). How to treat genetically modified crops correctly – reflection on the 'Golden Rice' trial. *Chinese Medical Ethics*, 26, 14–16.

Tian, X., Yuan H., and Ouyang, D. (2013). Reflections on the Golden Rice controversy: proper enforcement of ethics review through rationality and moral standards. *Chinese Medical Ethics*, 26, 17–20.

TIME (1959). Foreign news: facing starvation. *TIME Magazine*, 4 May. http://content.time.com/time/magazine/article/0,9171,892492,00.html.

TIME (1960). India: challenging Malthus. *TIME Magazine*, 9 May. http://content.time.com/time/subscriber/article/0,33009,897459,00.html.

TIME (1965). India: the threat of famine. *TIME Magazine*, 3 December. http://content.time.com/time/subscriber/printout/0,8816,842253,00.html#.

Times of India (2005). AIIMS claims cutting edge stem cell study. *Times of India*, 23 March. https://timesofindia.indiatimes.com/home/science/AIIMS-claims-cutting-edge-stem-cell-study/articleshow/1059744.cms.

Timmermans, S., and Epstein, S. (2010). A world of standards but not a standard world: toward a sociology of standards and standardization. *Annual Review of Sociology* 36(1), 69–89.

Tiwari, R. (2019). Each minister must have a 5-year plan: PM Modi to cabinet. *Indian Express*, 6 June. https://indianexpress.com/article/india/each-minister-must-have-a-5-yr-plan-pm-modi-to-cabinet-5767504/.

Tiwari, S., and Raman, S. (2014). Governing stem cell therapy in India: regulatory vacuum or jurisdictional ambiguity? *New Genetics and Society*, 33(4), 413–33.

Todhunter, C. (2014). Moily's parting shot. *Deccan Herald*, 6 March. www.deccanherald.com/content/389992/moilys-parting-shot.html.

Togocareer (2021). 2020 Chinese overseas returns employment survey. *Southern Finance Omnimedia Corp*, 29 January. https://m.sfccn.com/article/20210129/herald/ODA4LTM1MDgwOQ==.html.

Translational Medicine (2018). Shao Feng: youngest academian, overseas return with pure passion for science. *Sohu*, 17 October. www.sohu.com/a/260099993_183834.

Traweek, S. (1996). Kokusaika, Gaistsu and Bachigai: Japanese physcists' struggles for moving into the international political economy of science. In L. Nader (ed.), *Naked Science: Anthropological Inquiry into Boundaries, Power, and Knowledge*. London: Routledge.

Trines, S. (2019). New benefactors? How China and India are influencing education in Africa. World Education News and Reviews. 30 April 2019. https://wenr.wes.org/2019/04/how-china-and-india-are-influencing-education-in-africa.

Tuan, Y. (1977 [2001]). *Space and Place: The Perspective of Experience*. Minneapolis, MN: University of Minnesota Press.

UK-House of Lords (2012). Science and Technology Select Committee Regenerative Medicine: Oral and Written evidence. UK Parliament, 6 November 2012. www.parliament.uk/globalassets/documents/lords-committees/science-technology/RegenerativeMedicine/RegenMed.pdf.

UNESCO (United Nations Educational Scientific and Cultural Organization) (2008). *Asia Pacific Perspectives on Biotechnology and Bioethics*. Bangkok: UNESCO Bangkok.

Valkenburg, G. (2020). Consensus or contestation: reflections on governance of innovation in a context of heterogeneous knowledge. *Science, Technology & Society*, 25, 341–56.

Van Noorden, R. (2015). India by the numbers. *Nature*, 521, 142–3.

Van Noorden, R. (2020). Open-access Plan S to allow publishing in any journal. *Nature*, 16 July. https://doi.org/10.1038/d41586-020-02134-6.

Varma, S. (2011). Bhopal gas victims now turn guinea pigs. *Times of India*, 24 February. http://articles.timesofindia.indiatimes.com/2011-02-24/india/28627612_1_gas-victims-bhopal-memorial-hospital-gas-disaster.

Vemuganti, G. K., Sangwan, V. S., and Rao, G. N. (2007). The promise of stem cell therapy for eye disorders. *Clinical and Experimental Optometry*, 90, 315–16.

Verbeek, B., and Zalove, A. (2017). Populism and foreign policy. In C. R. Kaltwasser, P. Taggart, P. O. Espejo, and P. Ostiguy (eds), *The Oxford Handbook of Populism*. Oxford: Oxford University Press.

Verma, R. (2020). China's 'mask diplomacy' to change the COVID-19 narrative in Europe. *Asia Europe Journal*, 18, 205–9.

Vieira, M. A., and Alden, C. (2011). India, Brazil, and South Africa (IBSA): South–South cooperation and the paradox of regional leadership. *Global Governance*, 17, 507–28.

Wahlberg, A. (2018). Exposed biologies and the banking of reproductive vitality in China. *Science, Technology & Society*, 23, 307–23.

Walsh, C. E. (2018). The decolonial for: resurgences, shifts, and movements. In W. D. Mignolo and C. E. Walsh (eds) *On Decoloniality: Concepts, Analytics, Praxis*. Durham, NC: Duke University Press.

Walsh F. (2008). UK's first hybrid embryos created. BBC, 1 April. http://news.bbc.co.uk/1/hi/health/7323298.stm.

Walsh, L. (2013). *Scientists as Prophets: A Rhetorical Genealogy*. Oxford: Oxford University Press.

Wang, J., and Feng, Y. (2018). Research on India's science technology and innovation plan system from 1st FYP to 12th FYP. *Science and Technology Management Research* (in Chinese), 20, 30–9.

Wang, Q., Tang, L., and Li, H. (2015). Return migration of the highly-skilled in higher education institutions: a Chinese university case. *Population Space and Place*, 21, 771–87.

Wang, V., Qin, A., and Wee, S-L. (2020). Coronavirus survivors want answers, and China is silencing them. *New York Times*, 4 May. www.nytimes.com/2020/05/04/world/asia/china-coronavirus-answers.html?action=click&module=Spotlight&pgtype=Homepage.

Wang, Y-G. (2003).'Chinese ethical views on embryo stem (ES) cell research. In S. Song, and Y. Koo (eds), *Asian Bioethics in the 21st Century*. Bangkok: Eubios Ethics Institute.

Wang, Z. (2015) The Chinese developmental state during the Cold War: the making of the 1956 twelve-year science and technology plan. *History and Technology*, 31, 180–205.

Waters, J. (2019). Stem cell miracles transformed our lives: patients reveal how they have reaped remarkable benefits of medical revolution that may be as significant as antibiotics or organ transplants. *Daily Mail*, 11 November. www.dailymail.co.uk/health/article-7674229/Patients-reveal-reaped-remarkable-benefits-stem-cells.html.

Waters, R. (2015). Investor rush to artificial intelligence is real deal. *Financial Times*, 4 February. www.ft.com/content/019b3702-92a2-11e4-a1fd-00144feabdc0#axzz3Ny5kj89q.

Watts, J. (2004). I don't know how it works. *Guardian*, 1 December. www.guardian.co.uk/education/2004/dec/01/highereducation.uk1.

Watts, S. (2006). Stem cell treatment warning. BBC, 30 August. http://news.bbc.co.uk/1/hi/programmes/newsnight/5299306.stm.

WCSLRT (World Conference on Science Literacy Round Table) (2019). 2019 Memorandum of the World Conference on Science Literacy Round Table. World Conference on Science Literacy, 16 October. www.wcsl.org.cn/memorandum.

Wellcome Trust (1999). Human genome will be defined by spring. Wellcome Trust. www.genome.gov/10002109/1999-release-human-genome-will-be-defined-by-spring.

Wen, C. (2019). A housekeeper for research ethics. *China Science Daily*, 24 July. http://news.sciencenet.cn/htmlnews/2019/7/428834.shtm.

Wen, Y., and Lipes, J. (2016). Shandong's vaccine panic. *Radio Free Asia*. www.rfa.org/english/news/special/vaccinecrisis/report03.html.

WHO (World Health Organization) (2019). Ten Threats to Global Health in 2019. www.who.int/news-room/spotlight/ten-threats-to-global-health-in-20195.

Wilsdon, J., and Keeley, J. (2007). *China: The Next Science Superpower?* London: Demos.

Wilsdon, J., and Wills, R. (eds) (2004). *See-through Science: Why Public Engagement Needs to Move Upstream*. London: Demos.

WIPO (World Intellectual Property Organisation) (2020). World Intellectual Property Indicators Report. WIPO, Geneva, 7 December. www.wipo.int/pressroom/en/articles/2020/article_0027.html.

Wiyeh, A. B., Cooper, S., Nnaji, C. A., and Wiysonge, C. S. (2018). Vaccine hesitancy 'outbreaks': using epidemiological modeling of the spread of ideas to understand the effects of vaccine related events on vaccine hesitancy. *Expert Review of Vaccines*, 17, 1063–70.

Wouters, O. J., et al. (2021). Challenges in ensuring global access to COVID-19 vaccines: production, affordability, allocation, and deployment. *Lancet*, 397, 1023–34.

Wu, G., and Qiu, H. (2012). Popular science publishing in contemporary China. *Public Understanding of Science*, 22, 521–9.

Wu, J. (2021). The Majority of Chinese Hold Positive Attitude Towards Science and Technology. Xinhua, 3 January. www.liguo.cc/8/289075.html.

Wynne, B. (1980). Risk, technology and trust: on the social treatment of uncertainty. In J. Conrad (ed.), *Society, Technology and Risk*. London: Arnold.

Wynne, B. (2001). Creating public alienation: expert cultures of risk and ethics on GMOs. *Science as Culture* 10(4), 445–8.

Xiao, X.-J. (2004). The absence of 'ethics' in 'ethical guidelines'. *Scientific Times* (Kexue Shibao), 23 July.

Xinhua (2013). President Xi proposes Silk Road economic belt. *China Daily*, 7 September. www.chinadaily.com.cn/china/2013xivisitcenterasia/2013-09/07/content_16951811.htm.

Xinhua (2017). China's genomics company BGI makes stock market debut. Xinhua, 14 July. www.chinadaily.com.cn/business/2017-07/14/content_30117466.htm.

Xinhua (2021). The Fourteenth Five-Year-Plan of Social-Economic Development and the Outline of 2035 Vision of the People's Republic of China. Xinhua, 13 March. www.gov.cn/xinwen/2021-03/13/content_5592681.htm.

Xiong, L. (2021). Investigate Harvard's poaching of Chinese genes, experience the contest with capital. Netease News 3 January. www.163.com/dy/article/FVDE23IC0514C63D.html.

Xu, Q. (2010). Qiu Renzong: opening the door of bioethics for China. Wenhui Bao, 11 January. http://news.sciencenet.cn/htmlnews/2010/1/227070.shtm.

Yair, G. (2019). Culture counts more than money: Israeli critiques of German science. *Social Studies of Science*, 49, 898–918.

Yan, Z. (2017). MOF spokeperson on BRI being incorporated into the Party constitution: reflection of determination and confidence. Xinhua, 26 October. www.xinhuanet.com/world/2017-10/26/c_1121862497.htm.

Yang, R., Penders, B., and Horstman, K. (2020). Vaccine hesitancy in China: a qualitative study of stakeholders' perspectives. *Vaccines* (Basel), 8(4), 650.

Yang, Y. (2021). Wuhan mission unlikely to settle charged debate on virus origins. *Financial Times*, 9 February. www.ft.com/content/f07b6aa9-0746-470e-a04a-d6bba47a0a13?fbclid=IwAR0seyo1147mjSPONfgw-qo8bvynom-qEUBYpfjqsmWd_kng5qt1Y44J5Yw.

Ye, J. (2013). Foxconn of Biomedicine? The ruthless ride of 'bandit' entrepreneurs. *Ebiotrade*, 23 April. www.ebiotrade.com/newsf/2013-4/2013422144120580.htm.

Yee, A. (2012). Regulation failing to keep up with India's trials boom. *Lancet*, 379, 397–8.

Zahariadis, N. (2003). *Ambiguity and Choice in Public Policy: Political Decision Making in Modern Democracies*. Washington, DC: Georgetown University Press.

Zhai, X., Lei, R., Zhu, W., and Qiu, R. (2019). Chinese Bioethicists Respond to the Case of He Jiankui. *The Hastings Centre*, 7 February. www.thehastingscenter.org/chinese-bioethicists-respond-case-jiankui/.

Zhang, H. (2018). New mid-to long- term science and technology plan initiated, need for more autonomy given to researchers. 21st Century Business Herald, 12 December. http://epaper.21jingji.com/html/2018-12/12/content_98225.htm.

Zhang, H. (2021). Chinese vaccines slower than Pfizer in WHO validation, but 'intl orders unaffected'. *Global Times*, 8 January. www.globaltimes.cn/page/202101/1212261.shtml.

Zhang, J. Y. (2010). The cosmopolitanization of science: experience from China's stem cell scientists. *Soziale Welt* (special issue in English), 61, 255–74.

Zhang, J. Y. (2011). Scientific institutions and effective governance: a case study of Chinese stem cell research. *New Genetics and Society*, 30(2), 193–207.

Zhang, J. Y. (2012a). *The Cosmopolitanization of Science: Stem Cell Governance in China*. Basingstoke: Palgrave Macmillan.

Zhang, J. Y. (2012b) The art of trans-boundary governance of synthetic biology. *Systems and Synthetic Biology*, 7, 107–114.

Zhang, J. Y. (2014). Does China offer a new paradigm for doing science? In D. Kerr (ed.), *China's Many Dreams: Comparative Perspectives on China's Search for National Rejuvenation*. Basingstoke: Palgrave Macmillan.

Zhang, J. Y. (2015). The 'credibility paradox' in China's science communication: views from scientific practitioners. *Public Understanding of Science*, 24(8), 913–27.

Zhang, J. Y. (2017). Lost in translation? Accountability and governance of clinical stem cell research in China. *Regenerative Medicine*, 12(6), 647–56.

Zhang, J. Y. (2018a). How to be modern? The social negotiation of 'good food' in contemporary China. *Sociology*, 52(1), 150–65.

Zhang, J. Y. (2018b). *Governing Scientific Accountability in China. Final Report of the ESRC Research Project*. Canterbury: GSA-China.

Zhang, J. Y. (2021). Nurture science solidarity to keep nationalism in check. *Nature*, 591, 529.

Zhang, J. Y. (2022). The dual role of risk in mitigating socio-political inequality: a case study on the cosmopolitanization of science. In D. Curran (ed.), *Handbook on Risk and Inequality*. Cheltenham: Edward Elgar Publishing.

Zhang, J. Y., and Barr, M. (2013). *Green Politics in China: Environmental Governance and State–Society Relations*, London: Pluto Press.

Zhang, J. Y., and Barr, M. (2021). Harmoniously denied: COVID-19 and the latent effects of censorship. *Surveillance and Society*.

Zhang, L-R., and Chen, Y-Y. (2001). Major breakthrough for therapeutic cloning: Sun Yat-sen Medical University cloned more than 100 human embryos using new technology. *People's Daily* (Southern China News), 7 September. www.people.com.cn/GB/paper49/4169/485725.html.

Zhang, Q. (2020). Analysis of India's science and innovation capacity and inspirations for China. *Science and Technology of China*, 8, 27–32.

Zhao, K. (2020). The coronavirus story is too big for China to spin. *New York Times*, 14 February. www.nytimes.com/2020/02/14/opinion/china-coronavirus-social-media.html.

Zhao, W., and Chen, L. (2015). Experience and inspirations from India's Inclusive innovation. *Science & Technology Progress and Policy* (in Chinese), 32, 1–5.

Zhu, Y. (2016). First China-India Science, Technology and Innovation Collaborative Research Forum Held. Chinese Academy of Science and Technology for Development, 26 January. http://2015.casted.org.cn/web/index.php?ChannelID=7&NewsID=6354.

Zhu, Z., and Gong, X. (2008). Basic research: its impact on China's future. *Technology in Society*, 30, 293–8.

Zhu, Y., and Wu, J. (2006). A Chinese doctor's transnational IP struggle. 7 July, Xinhuanet. http://news3.xinhuanet.com/newscenter/2006-07/07/content_4806688.htm.

Ziman, J. (1978 [1996]). *Reliable Knowledge: An Exploration of the Grounds for Belief in Science*. Cambridge: Cambridge University Press.

Zimmer, K. (2018). CRISPR scientists slam methods used on gene-edited babies. *The Scientist*, 4 December. www.the-scientist.com/news-opinion/crispr-scientists-slam-methods-used-on-gene-edited-babies-65167.

Zweig, D., and Chen, C. (1995). *China's Brain Drain to the United States: Views of Overseas Chinese Students and Scholars in the 1990s (China Research Monograph Series)*. Berkeley, CA: Institute for East Asian Studies.

Zweig, D., Fung, C. S., and Han, D. (2008). Redefining the brain drain: China's 'diaspora option'. *Science, Technology & Society*, 13, 1–33.

Websites

Belt and Road Life Sciences Economy Alliance: www.brlsea.org/programme.

CanSino Nature Advertisement: www.nature.com/articles/d42473-018-00219-5.

Cell Transplantation: https://journals.sagepub.com/metrics/cll.

Genetic Engineering Appraisal Committee (GEAC), India: www.geacindia.gov.in.

International Association of Neurorestoratology: www.ianr.org.cn.

Journal of Neurorestoratology: http://jnr.tsinghuajournals.com/EN/2324-2426/home.shtml.

National Distribution of Gross Domestic Product (GDP): https://statista.com.

National Natural Science Foundation of China: http://nsfc.gov.cn.

Nutech MediWorld: www.nutechmediworld.in.

Peter Daszak tweets:
https://twitter.com/PeterDaszak/status/1360551108565999619.
https://twitter.com/PeterDaszak/status/1360591411469516801.

Thea Fischer tweet: https://twitter.com/TheaKFischer/status/1360590441817772034.

Traditional Knowledge Digital Library: www.tkdl.res.in/.

WHO COVAX: www.who.int/initiatives/act-accelerator/covax.

Index